Plant Improvement
and Somatic Cell Genetics

Academic Press Rapid Manuscript Reproduction

Based on two symposia held at the XIIIth International Botanical
Congress, Sydney, Australia: Cell Culture and Somatic Cell Genetics
in Plant Biology and Frontiers in Plant Breeding, held August 21–28,
1981.

Plant Improvement and Somatic Cell Genetics

Edited by

Indra K. Vasil

Department of Botany
University of Florida
Gainesville, Florida

William R. Scowcroft

Division of Plant Industry
CSIRO
Canberra City, Australia

Kenneth J. Frey

Department of Agronomy
Iowa State University
Ames, Iowa

 ACADEMIC PRESS, INC.

(Harcourt Brace Jovanovich, Publishers)

Orlando   San Diego   San Francisco   New York
London   Toronto   Montreal   Sydney   Tokyo

ACADEMIC PRESS, INC.
Orlando, Florida 32887

United Kingdom Edition published by
ACADEMIC PRESS, INC. (LONDON) LTD.
24/28 Oval Road, London NW1 7DX

Library of Congress Cataloging in Publication Data
Main entry under title:

Plant improvement and somatic cell genetics.

   Bibliography: p.
   Includes index.
   1. Plant-breeding. 2. Plant genetic engineering.
3. Plant cell culture. I. Vasil, I. K. II. Scowcroft,
William R. III. Frey, Kenneth J.
SB123.P563 1982   631.5'3   82-11467
ISBN 0-12-714980-5

PRINTED IN THE UNITED STATES OF AMERICA

84 85 86 87   9 8 7 6 5 4 3 2

Contents

Contributors

Numbers in parentheses indicate the pages on which the authors' contributions begin.

William L. Brown (29), *Pioneer Hi-Bred International, Inc., 1206 Mulberry Street, Des Moines, Iowa 50308*

Chih-Ching Chu (129), *Institute of Botany, Academia Sinica, 141 Hsi Chih Men Wai Ta Chie, Beijing, China*

Agnes Cséplö (221), *Institute of Plant Physiology, Biological Research Centre, Hungarian Academy of Sciences, 6701 Szeged, Hungary*

Henri De Greve (255), *Laboratory GEVI, Vrije Universiteit Brussel, St.-Genesius-Rode, Belgium*

Ann Depicker (255), *Laboratory of Genetics, Rijsuniversiteit Gent, Gent, Belgium*

Patrick Dhaese (255), *Laboratory of Genetics, Rijsuniversiteit Gent, Gent, Belgium*

Trinh Manh Dung*(221), *Institute of Plant Physiology, Biological Research Centre, Hungarian Academy of Sciences, 6701 Szeged, Hungary*

Richard B. Flavell (277), *Plant Breeding Institute, Maris Lane, Trumpington, Cambridge, CB2 2LQ, United Kingdom*

Kenneth J. Frey (43), *Department of Agronomy, Iowa State University, Ames, Iowa 50011*

Esra Galun (205), *Department of Plant Genetics, The Weizmann Institute of Science, Rehovot, Israel*

Jean-Pierre Hernalsteens (255), *Laboratory GEVI, Vrije Universiteit Brussel, St.-Genesius-Rode, Belgium*

Marcelle Holsters (255), *Laboratory of Genetics, Rijsuniversiteit Gent, Gent, Belgium*

Gabriella Lázár (221), *Institute of Plant Physiology, Biological Research Centre, Hungarian Academy of Sciences, 6701 Szeged, Hungary*

*Present address: *Institute of Experimental Botany, Ho Chi Minh City, Vietnam.*

.. J. Larkin (159), *Division of Plant Industry, CSIRO, Canberra City, Australia*

Jan Leemans (255), *Laboratory GEVI, Vrije Universiteit Brussel, St.-Genesius-Rode, Belgium*

Pál Maliga (221), *Institute of Plant Physiology, Biological Research Centre, Hungarian Academy of Sciences, 6701 Szeged, Hungary*

László Marton (221), *Institute of Plant Physiology, Biological Research Centre, Hungarian Academy of Sciences, 6701 Szeged, Hungary*

Peter Medgyesy (221), *Institute of Plant Physiology, Biological Research Centre, Hungarian Academy of Sciences, 6701 Szeged, Hungary*

László Menczel (221), *Institute of Plant Physiology, Biological Research Centre, Hungarian Academy of Sciences, 6701 Szeged, Hungary*

Ferenc Nagy (221), *Institute of Plant Physiology, Biological Research Centre, Hungarian Academy of Sciences, 6701 Szeged, Hungary*

Leon Otten (255), *Max-Planck-Institut für Züchtungsforschung, Köln, Federal Republic of Germany*

F. N. Ponnamperuma (73), *International Rice Research Institute, Manila, Philippines*

Charles M. Rick (1), *Department of Vegetable Crops, University of California, Davis, California 95616*

Jeff Schell* (255), *Laboratory of Genetics, Rijsuniversiteit Gent, Gent, Belgium*

Otto Schieder (239), *Max-Planck-Institut für Züchtungsforschung (Erwin-Baur-Institut), D-5000 Köln 30 (Vogelsang), Federal Republic of Germany*

Gudrun Schröder (255), *Max-Planck-Institut für Züchtungsforschung, D-5000 Köln 30, Federal Republic of Germany*

Jo Schröder (255), *Max-Planck-Institut für Züchtungsforschung, D-5000 Köln 30, Federal Republic of Germany*

W. R. Scowcroft (159), *Division of Plant Industry, CSIRO, Canberra City, Australia*

Vladimir Sidorov† (221), *Institute of Plant Physiology, Biological Research Centre, Hungarian Academy of Sciences, 6701 Szeged, Hungary*

Jacqueline James Sprague (99), *Centro Internationale de Mejoramiento de Maizy Trigo, El Batan, Mexico*

Marc Van Montagu (255), *Laboratory of Genetics, Rijsuniversiteit Gent, Gent, Belgium*

Indra K. Vasil (179), *Department of Botany, University of Florida, Gainesville, Florida 32611*

Isabelle M. Verry (119), *Grasslands Division, D. S. I. R., Palmerston North, New Zealand*

Elizabeth G. Williams (119), *Grasslands Division, D. S. I. R., Palmerston North, New Zealand*

*Present address: *Max-Planck-Institut für Züchtungsforschung, D-5000 Köln 30, Federal Republic of Germany.*

†Present address: *Institute of Botany, Kiev, USSR.*

Warren M. Williams (119), *Grasslands Division, D. S. I. R., Palmerston North, New Zealand*

Lothar Willmitzer (255), *Max-Planck-Institut für Züchtungsforschung, D-5000 Köln 30, Federal Republic of Germany*

Patricia Zambryski (255), *Laboratory of Genetics, Rijsuniversiteit Gent, Gent, Belgium*

Preface

Two symposia related to plant improvement and somatic cell genetics were held during the XIII International Botanical Congress at Sydney, Australia (August 21–28, 1981). One symposium, "Frontiers in Plant Breeding," was organized by Kenneth J. Frey and William R. Scowcroft and was chaired by Sir Otto Frankel. A related symposium, "Cell Culture and Somatic Cell Genetics in Plant Biology," was organized by Indra K. Vasil and William R. Scowcroft and was chaired by Indra K. Vasil. A common and continuing theme of the two symposia was plant improvement. This volume includes all but one of the papers presented at these two symposia. Several of the chapters illustrate convincingly the impressive power and the remarkable success of plant breeding techniques in crop improvement. The chapters on cell culture, somatic cell genetics, and molecular biology bring forth the drama and potential of these novel techniques, which in future may help to overcome some of the limitations of conventional plant breeding by providing a broader genetic base for gene flow and superior plants.

The value of these collected papers lies in the integration achieved at the symposia in Sydney and in this volume between conventional plant breeding and cell culture and somatic cell genetics. Leading scientists from many countries examined the potentials as well as the limitations of plant breeding and cellular and molecular techniques in plant improvement. The most important lesson to be learned from this presentation is that only a close combination of conventional plant breeding and genetics with the emerging techniques of cellular and molecular biology will allow for the utilization of a broad genetic base for the creation of a range of genetic diversity hitherto unattainable. This in time may lead to the production and selection of superior and improved plants.

We would like to acknowledge the financial support received by the participants from their home institutions and various other sources. Both symposia were also supported by the Australian Academy of Sciences, the Genetics Society of Australia, and the Australian Wheat Industry Research Council.

Indra K. Vasil
William R. Scowcroft
Kenneth J. Frey

CHAPTER 1

THE POTENTIAL OF EXOTIC GERMPLASM
FOR TOMATO IMPROVEMENT[1]

Charles M. Rick

Department of Vegetable Crops
University of California
Davis, California

I. DISEASE RESISTANCE

A truly impressive record has been established by workers
in search of high-level resistance and in its incorporation
into acceptable cultivars of tomatoes. This category is con-
strued in the broad sense, including bacteria, fungi, viruses,
nematodes, and physiological disorders. In many areas, primarily temperate, the disease problem has been effectively and
economically solved by means of inherited resistance. Combined resistance to several pathogens has been achieved to the
extent that cultivars with resistance to three and four major
diseases are now commonplace and larger combinations not unknown. The hybrid cultivar Floramerica, for example, pos-
esses monogenic resistance against five diseases and tolerance
of five others; additionally, it is free of three physiological disorders (Crill et al., 1977). In some instances (e.g.,
nematodes), resistance to several species and many biotypes of
the parasites has been effectively achieved by introduction of
the single *Mi* gene derived from *L. peruvianum* (Gilbert and
McGuire, 1956); in many others, including resistance to
Cladosporium, curly top, and spotted wilt, adequate resistance
has required the combined action of a battery of genes. The
great majority of these responsible genes have been derived

[1] *Bibliographic research supported by NSF Grant DEB 80-05542*

1

from wild forms, a fact underscored by the absence of cultivars with appreciable disease resistance prior to Porte and
Wellman's (1941) introduction of a fusarium-resistant cultivar; previously, tomato breeders leaned entirely on the
impoverished pool of germplasm available in European and North
American cultivated stocks. Satisfactory resistance to at
least 24 diseases has been found in wild forms (Table 1.1); in
14 of these cases the pertinent genes have been bred into productive cultivars; and for each of at least six diseases,
literally dozens of resistant cultivars have been introduced.

II. ARTHROPOD RESISTANCE

Here again, sources of resistance offer hope that destruc-
tive insects and spider mites might be controlled by resistance or tolerance derived from wild or exotic sources of the
tomato. Also, as in the case of disease resistance, high-level
resistance is rarely encountered in cultivated forms of *L.
esculentum*. The record (Table 1.2) reveals resistance to 16
such pests, judged to be high level or near immunity. It is
remarkable that *L. hirsutum* provides high resistance to 14 of
these parasites and is the only known source of resistance to
nine. Also noteworthy is the fact that the precise nature of
resistance in this species is different against different
pests. Thus, Williams et al. (1980) have ascertained that the
sesquiterpenoid, 2-tridecanone in glandular trichome exudate
is the toxic factor that is antibiotic to *Manduca sexta*;
Patterson et al. (1975) have implicated two sesquiterpenoids
from foliage in the resistance of this species to *Tetranychus
urticae*; while high trichome density has been identified by
Gentile et al. (1968) as the responsible factor in its resistance to *Trialeurodes vaporarium*.
In November 1980 the writer was privileged to visit a
field test conducted in Yuto (Salta), Argentina for resistance
of tomato lines to the lepidopterous leafminer *Scrobipalpula
absoluta*, a severe pest of tomatoes there. Local experts have
found that control by insecticides has not been satisfactory
because biotypes of this insect resistant to the chemicals are
generated rapidly by selection. Under conditions of this test,
standard tomato cultivars were uniformly defoliated and eventu-
ally killed, whereas plants of *L. hirsutum* f. *glabratum* were
literally unscathed, and the F_1 hybrid evinced an intermediate
level of resistance (Fig. 1.1; Gilardón and Benavent, 1981).
The dramatic capacity of *L. hirsutum* to resist attack of a wide
range of arthropod predators is concordant with observations of
this species in its native habitat. In visiting many habitats

in Ecuador and Peru, we have consistently observed the free-
dom, particularly of foliage, of *L. hirsutum* from arthropod
attack (Rick, 1973).

Notwithstanding the astonishing resistance that is available in tomato germplasm sources, progress in breeding resistant cultivars has been slow. The recentness of research in
this area, and many technical difficulties of evaluating insect resistance account to some extent for the lag of productivity in these efforts.

III.   STRESS TOLERANCE

Since this topic was reviewed eight years ago (Rick, 1973),
only new developments and their implications for tomato improvement will be considered in this section. As research in
tomato stress tolerance has been largely limited to the past
decade, sufficient time has not elapsed for generation of
stress-tolerant cultivars; hence assessments of the prospects
for future progress would be premature. Therefore, the work
reviewed has been concerned to a much greater extent with
evaluating existing germplasm. Progress in genetic aspects
is bound to be slow for several reasons. Programs cannot
develop until appropriate methodology has been perfected for
screening the breeding populations. Further, the high environmental variance typical of such characters reduces selection efficiency. Another factor is that epistasis may occur
between the introduced genes and the *esculentum* genetic background. Economic incentives must also be considered: Although the importance of resistance to salinity, temperature
extremes, drought, and other environmental stresses has been
widely publicized, growers will not move into marginal land
subject to these stresses unless it is their last resort, a
situation that contrasts with the motivation for plant breeding engendered by outbreaks of new diseases.

Evaluating comparative fruit setting capacity under various
stress conditions is complicated by outcrossing in many
tomato species. Self-incompatibility enforces cross-pollination in *L. chilense*, *L. hirsutum*, *L. peruvianum*, and *S. pennellii*; furthermore, showiness of flowers, exserted stigmas
and other factors promote outcrossing in the facultatively
allogamous accessions of *L. pimpinellifolium*, *L. chmielewskii*,
L. hirsutum, and *S. pennellii*. When grown in the field outside their native habitats and particularly in enclosures,
these taxa require artificial pollination for fruit set.
Testing environmental effects, temperature extremes, for example, therefore becomes particularly difficult since the

Table 1.1. Disease resistance in wild tomato species.

Disease	Source of resistance? Resistant host	Reference[a]
BACTERIA		
*Corynebacterium michiganense (Bacterial canker)	L. hirsutum L. pimpinellifolium L. hirsutum f. glabratum L. peruvianum L. peruvianum (glandulosum) L. peruvianum var. humifusum	deJong and Honma, 1976 Kuruyama and Kuruyasu, 1974 Laterrot et al., 1978
*Pseudomonas solanacearum (Bacterial wilt)	L. pimpinellifolium	Acosta et al., 1965 Villareal and Lai, 1978
Pseudomonas tomato (Bacterial speck)	L. pimpinellifolium	Laterrot and Rat, 1981
Xanthomonas vesicatoria (Bacterial spot)	L. esculentum cerasiforme	Mathew and Patel, 1975
FUNGI		
Alternaria solani (Collar rot)	L. pimpinellifolium L. hirsutum L. peruvianum var. humifusum	Andrus et al., 1942 Hódosy, 1975
*Cladosporium fulvum (Leaf mold)	L. esculentum var. cerasiforme	Kerr and Bailey, 1964
*Colletotrichium coccodes (Fruit anthracnose)	L. esculentum var. cerasiforme	Stevenson et al., 1978

4

Pathogen (Disease)	Lycopersicon species	Reference
Corynespora cassiicola (Target leaf spot)	L. pimpinellifolium	Bliss et al., 1973
Didymella lycopersici (Didymella canker)	L. hirsutum L. hirsutum f. glabratum	Anon., 1971 van Steekelenburg, 1981
*Fusarium oxysporum (Fusarium wilt)	L. pimpinellifolium	Bohn and Tucker, 1940
*Phytophthora infestans (Late blight)	L. pimpinellifolium ancestry suspected	Gallegly and Marvel, 1955
*Pyrenochaeta lycopersici (Corky root)	L. Peruvianum (glandulosum) L. hirsutum f. glabratum	Hogenboom, 1970 Maxon-Smith, 1978
*Septoria lycopersici (Leaf spot)	L. hirsutum L. pimpinellifolium L. esculentum var. cerasiforme	Andrus and Reynard, 1945 Barksdale and Stoner, 1978
*Stemphylium solani (Grey leaf spot)	L. pimpinellifolium	Hendrix and Frazier
*Verticillium albo-atrum (Verticillium wilt)	L. esculentum var. cerasiforme	Schaible et al., 1951
Verticillium dahliae	L. peruvianum	Saccardo and Ancora, 1981
NEMATODES		
Globodera pallida (Potato cyst nematode)	L. hirsutum f. glabratum	Maxon-Smith, 1978
Heterodera schachtii (Sugarbeet nematode)	L. pimpinellifolium	Steele, 1977

Table 1.1. (continued)

Disease	Source of resistance? Resistant host	Reference[a]
Meloidogyne incognita (Root-knot nematode)	*L. peruvianum*	Gilbert and McGuire, 1956
VIRUSES		
CMV (Cucumber mosaic virus)	*L. peruvianum* *S. lycopersicides*	Jacquemond and Laterrot, 1981
*CTV (Curly top virus)	*L. peruvianum* var. *dentatum* *L. peruvianum* var. *humifusum*	Martin, 1970
*SWV (Spotted wilt virus)	*L. pimpinellifolium*	Finlay, 1953
*TMV (Tobacco mosaic virus)	*L. peruvianum*	Alexander, 1963
TYLCV (Tomato yellow leaf virus)	*L. pimpinellifolium* *L. cheesmanii* *L. hirsutum* *L. hirsutum* f. *glabratum* *L. peruvianum* *L. peruvianum* var. *humifusum*	Pilowsky and Cohen, 1974 Hassan et al., 1982

NON-PATHOGENIC DISORDERS

Blossom end-rot   Virtually all wild taxa   Casual observations of many
workers

Silvering   *L. cheesmanii*   Grimbly, 1981
L. hirsutum f. *glabratum*
S. pennellii

a *Only key references are cited.*

* *Resistance bred into cultivars.*

7

Table 1.2. Resistance in the tomato species to arthropod pests.

Pest	Resistant host	Reference[a]
Coleoptera		
Leptinotarsa decemlineata (Colorado potato beetle)	*L. hirsutum* f. *glabratum*	Sinden et al., 1978
Epitrix hirtipennis (Tobacco flea beetle)	*L. hirsutum* f. *glabratum*	Gentile and Stoner, 1968b
Diptera		
Liriomyza munda (Leafminer)	*L. hirsutum* f. *glabratum*	Webb et al, 1971
L. sativae (Vegetable leafminer)	*L. hirsutum* f. *glabratum*	Schuster et al., 1979
Homoptera		
Aphis caccivore	*L. hirsutum* f. *glabratum*	Kennedy and Yamamoto, 1979
A. gossypii	*L. hirsutum* f. *glabratum*	Williams et al., 1980
Macrosiphon euphorbia (Potato aphid)	*L. peruvianum, S. pennellii*	Gentile and Stoner, 1968a
Myzus persicae (Peach aphid)	*L. hirsutum* f. *glabratum*	Kennedy and Yamamoto, 1979
Trialeurodes vapororium (Greenhouse whitefly)	*L. hirsutum, S. pennellii* *L. pimpinellifolium, chilense*	Gentile et al., 1968 Hogenboom et al., 1974

Lepidoptera

Heliothis zea
(Tomato fruitworm)

L. hirsutum, L. h. f.
glabratum

Fery and Cuthbert, 1975

Keiferia lycopersicella
(Tomato pinworm)

L. cheesmanii f. minor,
L. hirsutum, L. peruvianum
(glandulosum)

Schuster, 1977

Manduca sexta
(Tobacco hornworm)

L. hirsutum f. glabratum

Kennedy and Henderson, 1978

Scrobipalpula absoluta
(Leafminer)

L. hirsutum f. glabratum

Gilardón and Benavent, 1981

Spodoptera exigua
(Sugarbeet armyworm)

L. pimpinellifolium, L.
esculentum var. cerasiforme

Juvik, 1980

Acarina

Tetranychus cinnibarinum
(Carmine spider mite)

L. hirsutum f. glabratum
S. Pennellii

Gentile et al., 1969

T. urticae
(Two-spotted spider mite)

L. hirsutum f. glabratum

Gentile et al., 1969

a Only key references are cited.

9

Figure 1.1. Field test for resistance to the lepidopterous leafminer *Scrobipalpula absoluta* at Yuto (Prov. Salta), Argentina. *Lycopersicon hirsutum* f. *glabratum* on left, vestiges of *L. esculentum* in center row, F_1 *L. esculentum* x *L. hirsutum* f. *glabratum* on right.

stress conditions *per se* might affect some phase of the pollination process. The potential to set fruit under adverse conditions consequently cannot be adequately tested in the accessions themselves but has to be evaluated in hybrid derivatives. Thus, at the present time, practically nothing is known about fruit setting capacity under adverse conditions in the wild taxa.

High Temperature

It is commonly but erroneously assumed that, since the tomato is native to tropical latitudes, it should produce well at high temperatures. The adverse effects of excessively high temperatures on tomato performance are felt on a large scale in many areas. A frequent disorder is poor fruit set.

A detailed review of research on the physiology and genetics of this high temperature response is not warranted here; instead, attention will be given to the definitive investigations of El Ahmadi and Stevens (1979a, b), which reveal

the nature of the effects on fruit setting and of genetic
variation for them. Tolerance to high temperatures was found
in five accessions, including three exotic cultivars. The
analysis of components of fruit set (male and female gamete
viability, pollen release, stigma exsertion) revealed that all
of these lines do not respond to the stress in the same fashion. Since none of the lines combined all of the characteristics needed for high temperature setting ability, the investigators opined that, if such a combination were attain-
able, it should be an ideal heat-tolerant tomato. In an extensive survey of many accessions, Villareal and Lai (1979)
observed superior performance of several accessions of *L.
pimpinellifolium* under high temperatures.

Fruit setting is not the only problem of fruit production
under high temperatures. Even if setting is satisfactory,
fruit color may be poor because temperatures of 32°C or higher
inhibit lycopene (but not β-carotene) synthesis (Tomes, 1963).

Low Temperatures

Adaptation of the tomato to warm climates is evident in
the sensitivity of plant parts, including fruits, to prolonged
exposure to temperatures below 10°C. Before this sensitivity
was appreciated, great losses were suffered by shippers,
handlers, and consumers from physiological breakdown of fruits
resulting from prolonged refrigeration. Surveys of *esculentum*
germplasm have not turned up any promising tolerance of such
chilling, but in testing other species, Patterson et al.
(1978) and Vallejos (1979) discovered that high-altitude
biotypes of *L. hirsutum* can withstand such treatment in contrast to the sensitivity of biotypes from lower elevations and
the vast majority of other tomato species accessions. The
same relationships were ascertained in seed germination at low
temperatures, a trait found to be inherited in complex fashion (Paull et al., 1979). Zamir et al. (1981) reported that
chilling-tolerant genotypes derived from *L. hirsutum* might be
selected in male gametes. Currently the transfer of this
trait to a tomato background is the subject of intense research activity with the objectives of thereby alleviating the
risk of physiological fruit damage, of extending tomato cultivation into cooler areas, and of reducing the energy costs of
tomato production in greenhouses.

The possibilities of exploiting tolerance of even lower
temperatures are being pursued by Robinson and Kowaleski
(1974) in hybrids with the frost-tolerant *S. lycopersicoides*,
which is endemic to altitudes of 3,000 m or more in southern

Peru. The severe barrier to gene exchange between the latter
and *L. esculentum* must be circumvented before this goal can
be achieved.

Salinity

The cultivated tomato is highly sensitive to saline conditions. As might be anticipated, the immense variety of habitats occupied by *Lycopersicon* species include some of saline
nature. Although salinity tolerance has been identified in
certain biotypes, the full potential of this useful germplasm
variant has not been tapped. Recent research has been dedi-
cated largely to biotypes of *L. cheesmanii* and the greenfruited species.

Whereas most accessions of the Galápagos species *L.
cheesmanii* grow in average soil conditions, occasional populations of f. *minor* are encountered in littoral habitats, where
higher levels of salinity can be anticipated from exposure to
salt water seepage and salt spray. An accession flourishing
under conditions of extreme exposure was collected and distributed to specialists in the physiology of salt tolerance
(Rick, 1973). Tests made by Rush and Epstein (1976) revealed
that this biotype would survive in full-strength sea water if
the hydroponic medium in which it was cultured was gradually
increased to that concentration; in contrast, plants of *L. esculentum* collapsed at half-strength sea water. They also determined that resistance was associated, not with restrictions
on sodium uptake or translocation, but with an ability of the
cells to cope with increased sodium levels. By hybridizing
this accession with *L. esculentum* and backcrossing to the
latter, Frederickson and Epstein (1975) detected an increase
in salt tolerance in the hybrid progenies above that of the
esculentum parent. Rush and Epstein (1980) determined that
the salt tolerant *L. cheesmanii* accumulates sodium in leaf and
petiole in amounts exceeding 20% of the total dry weight,
whereas in *L. esculentum* sodium tends to be excluded under
salt stress and is lethal at concentrations above 5%. They
also noted that hybrid progeny selected for salt tolerance
accumulate sodium freely as in the tolerant parent and concluded that foliar sodium can be used as a convenient index of
salt tolerance. Progress in the transfer of this trait to
horticultural tomatoes was reported by Epstein et al. (1980):
after repeated backcrosses, derivatives have been obtained
that will continue to grow and produce fruit when irrigated
with 70% sea water.

Salt tolerance was also detected by Tal (1971), not only in
L. cheesmanii but also in *L. peruvianum*. In further studies

in *L. peruvianum* as well as in *S. pennellii*, Tal and his colleagues (Tal et al., 1978; Dehan and Tal, 1978; Tal et al.,
1979; Tal and Katz, 1980; Katz and Tal, 1980; Rosen and Tal,
1981) approached the problem in a different fashion from that
of Epstein and co-workers. Most of the research by the Tal
team has been devoted to responses of protoplast and callus
cultures to increased salinity levels. They determined that
growth of such cell cultures under salt stress is frequently
associated with accumulation of proline. The potential of
these additional sources of salt tolerance is revealed by
their investigations.

Drought

Even though germplasm banks of the tomato species include
accessions with known increased ability to withstand drought,
remarkably little research has been invested in this aspect.
Richards et al. (1979) report moisture stress tolerance in
two accessions of *L. esculentum* var. *cerasiforme*. The prime
example of drought resistance, however, is found in *S. pennellii*. All collections of this species have been made in
arid situations, and all tested accessions have exhibited a
unique ability of the leaves to retain moisture (Rick, 1973,
unpublished; Yu, 1972). We have been able to maintain this
character through several backcross generations, but additional research is needed to assess its potentialities for
imparting drought tolerance in the derived lines.

Excessive Moisture

Tomato production in the wet tropics is frequently limited
by excessive rainfall and consequent flooding. Since most of
the *Lycopersicon* species are native to arid or semi-arid conditions, it is not surprising that the cultivated tomato does
not tolerate waterlogging and that surveys of world collections have not identified many accessions that can endure such
conditions. With autecology as a clue, workers logically have
sought tolerance in biotypes that thrive in the wet tropics.
Accordingly, Rebigan et al. (1977) discovered tolerance in
two accessions from such conditions: Nagcarlan, a primitive
cultivar from the Philippines, and LA 1421, a feral member of
the *cerasiforme* complex from NE Ecuador. In the collection
site, the latter was found to survive heavy rains and high
water tables, conditions that decimated introduced cultivars
(Rick, 1973). In contrast to adverse effects on hundreds of
lines, neither of these two accessions was seriously affected

by excess water or poor drainage. They evidently possess
root characteristics that withstand such conditions and should
serve as the source of germplasm for breeding to adapt tomatoes to the wet tropics.

Mineral Utilization Efficiency

Surveys for efficiency in utilization of potassium and
nitrogen have been made within *L. esculentum*. One of two
highly efficient potassium utilizers selected by Makmur et al.
(1978) for studies of physiology and inheritance was a Peru-
vian accession of var. *cerasiforme*. Although no significant
differences were observed under non-stress conditions, this
line produced 79% more dry weight than inefficient lines when
the supply of potassium was restricted. The investigators
conclude that the differences in growth at 5 mg potassium/
plant are of "sufficient magnitude to permit selection of
superior individuals in segregating populations." Similar results were reported for efficiency of nitrogen utilization
(O'Sullivan et al., 1974). Under severe nitrogen stress an
accession of var. *cerasiforme* was much more efficient than
other lines in gaining dry weight. According to the investigators, the differences, the nature of genetic determination,
and economic need to improve fertilizer efficiency would
justify a program to breed improved nitrogen utilization into
acceptable cultivars.

IV. FRUIT QUALITY

The components of tomato fruit quality have been the subject of much recent investigation. In this field the reviews
by Stevens (1970a, 1974) are standard references, not only for
tomatoes but for vegetables in general. As expected, stan-
dards of quality vary according to uses, regions, and personal
preferences. The major dichotomy in fruit quality is in consumption as a fresh vs. processed vegetable: optimal characteristics differ for certain quality components, particularly
size, shape, firmness, flavor, and consistency. The role of
wild and exotic germplasm likewise varies with different
quality components.

Pigmentation

Major sources of useful genes are the redfruited species
L. esculentum and *L. pimpinellifolium*. Both have contributed
to intensification of color as evaluated internally and ex-
ternally. But greenfruited species have also contributed in
an unanticipated fashion: their genes can interact with those
of the cultivated parent to generate a type of interspecific
novel variation. Two examples can be cited. Lincoln and
Porter (1950) bred *L. esculentum* with *L. hirsutum* and selected
for modified fruit color in backcrosses to the former. They
discovered a monogenic segregation for orange vs. red fruits,
the dominant *B* gene from *hirsutum* determining orange color by
directing carotenoid synthesis toward β-carotene instead of
the normal red lycopene pigment. *B* exerts no detectable effect in the *hirsutum* parent. In the other case, a recombinant
with intensified lycopene synthesis was found in progenies of
L. esculentum x *L. chmielewskii* (Rick, 1974). Segregation in
backcrosses to the former parent indicated the interaction of
a single dominant *chmielewskii* gene *Ip* and the *esculentum*
background. Again, the presence of this gene could not be
detected in either parental species.

Soluble Solids

More attention has been paid to this component of quality
than to any other, because slight increases not only affect
flavor but also can effect great savings to the processing industry. Reducing sugars, which make up the bulk of soluble
solids, were found to be unusually abundant in fruits of *L.
chmielewskii*, and by successive backcrosses to *L. esculentum*
and selection, it was possible to derive large redfruited lines
with markedly higher refractometer readings (Rick, 1974).
The lines from this program are being used by other workers
for further improvement of tomato quality in both fresh-market
and processing tomatoes, whereas attempts with other sources
of high solids have not fared well, possibly because this
character suffers a high inverse relation with yields (Stevens
and Rudich, 1978). The physiological basis for this trait has
been investigated by Hewitt and Stevens (1981) and Hewitt et
al. (1982). They ascribe the high soluble solids level of a
chmielewskii derivative to two causes: (1) a lower fruit/leaf
ratio; (2) its fruits are stronger sinks for carbohydrates as
verified by the movement of labelled carbon from leaves. In a
study of the nature of the sink relationship, Dinar and
Stevens (1981) found that the leaf-fruit gradient was improved
by a more rapid sugar-starch conversion in developing fruits

of the *chmielewskii* derivative. This conversion is so consistent that selection for soluble solids can be done more
effectively by testing starch concentrations in 3-4 wk fruits.
Thus, curiosity of the investigators about the basic nature of
the difference between lines provided, among other benefits, a
payoff to the breeder to improve his selection efficiency.

Acidity

Tomato flavor depends to some extent on the sugar/acid
balance. Acidity also plays a critical role in the preservation of canned products: fruit with pH exceeding 4.4 is
considered less safe for standard processing methods. Thus a
cultivar with acidity approaching this level will be less desirable, for if grown under conditions favoring decreased
acidity, the risk of spoilage increases. Stevens (1972) sur-
veyed accessions and investigated the inheritance of different
levels of citric and malic acids. An accession of *L. pimpinellifolium* had the greatest titratable acidity, twice the
usual tomato level for citrate and thrice for malate. Monogenic inheritance of both traits simplifies the task of introgressing increased acidity.

Flavor Compounds

Other important flavor components are the 40-odd volatile
compounds identified in ripe tomatoes (Stevens, 1970a).
Stevens (1970b) investigated the contribution to flavor and
the inheritance of three principal volatiles and found inter-
varietal differences to be relatively simply inherited. In
another investigation, he detected an association between the
predominant polyene-carotene content and volatile compound
composition of several tested lines, suggesting that the
aromas could be influenced by the precursor polyene-carotenes
from which they were derived by oxidative degradation
(Stevens, 1970c). Thus the dominant neurosporene would yield
geranylacetone in cv. Jubilee and β-carotene, β-ionone in cv.
High Beta. Despite the low general popularity of non-red
tomatoes, the unique flavors of such cultivars are preferred
by certain consumers.

Five tomato cultivars and a breeding line were subjected to
detailed analysis of flavor components in relation or organo-
leptic evaluation by Stevens et al. (1977). Among their many
noteworthy results, of the many volatiles tested, three, including 2-isobutylthiazole, are the most important for tomato-
like flavor. Large differences in volatile concentration were

found between the lines, unusually high levels of 2-isobutyl-thiazole and another of these three volatiles, as well as of the majority of all tested volatiles, were detected in the aforementioned *chmielewskii* derivative (Rick, 1974).

Similar analysis apparently has not been extended to the wild species. Investigations can be expected to concentrate on cultivated sources, considering the complexities of analysis and rather revolting flavors (to human tastes) of the greenfruited species. The potentialities of wild sources, which could be surprisingly good, will not be known in the immediate future and will not be understood properly until responsible genes have been backcrossed into *esculentum* backgrounds.

Nutritional Factors

The primary nutritional value of the tomato lies in its content of vitamins and minerals. As indicated by Stevens (1974), although percentage composition is not high in the tomato, by virtue of the volume of fruit consumed by the public, the tomato ranks at the top of the list of all natural sources of these dietary components in the USA. The major constituent is vitamin C, which is extraordinarily variable in the tomato species. Yeager and Purinton (1946) exploited the very high content of *L. peruvianum* in a program of backcrossing and selection to produce cv. Hi-C. The cv. Doublerich with about twice the normal content of vitamin C was bred from the same cross by Schultz (1953).

The aforementioned research by Lincoln and Porter (1950) on the effects of *B* on β-carotene (provitamin A) content in *hirsutum* derivatives paved the way for breeding new cvs. such as Caro-Red (Tomes and Quackenbush, 1958) and others with much increased levels of this vitamin.

V. MISCELLANEOUS MORPHOLOGICAL AND PHYSIOLOGICAL CHARACTERS

Male Sterility

A novel type of male sterility has been reported in hybrid derivatives of *L. esculentum* x *L. parviflorum* (Kesicki, 1980). In this example the sterile anther tips are split and recurved, thereby exposing the stigma, and pollen release is delayed. Since its pollen is viable, this variant can be artificially self-pollinated and maintained in pure condition. Strongly modified cotyledons are part of the pleiotropic syndrome of

this male sterility. By virtue of these characteristics, this
line is effectively male-sterile; nearly all of the progeny
after cross-pollination are hybrids and the few selfed progeny
can be identified by their cotyledon shape. Since this male
sterility is not part of the parental phenotypes, it consti-
tutes another example of interspecific novel variation.

Jointless Pedicel

Two genes, j-2 and j-2^{in}, modifying the pedicel articula-
tion have been discovered in accessions of L. *cheesmanii*
(Rick, 1967). The former completely eliminates the articula-
tion, whilst the latter renders it non-functional. The former
was found in only a single population, but the latter is
rather widespread in the Galápagos archipelago. The j-2
allele has been incorporated in several cultivars and is being
exploited widely in the breeding of mechanically harvested
cultivars and in other situations in which vine retention of
fruits and/or complete elimination of the pedicel stub are important. The major defects in calyx structure and adhesion of
the fruit to the calyx that were associated with j-2 in the
early generations of the backcross transfer to L. *esculentum*
were eliminated by selection in later generations by several
workers, notably Dr. C. John of Heinz USA. An unusually thick
pericarp of certain accessions is another obviously useful
trait of L. *cheesmanii*.

Postharvest Physiology

Fruit ripening behavior of various tomato species was investigated by Grummet et al. (1981), who encountered a remarkable array of patterns differing *inter se* and from that of *esculentum* cultivars. As an example, fruits of L. *chilense* ab-
scise before ripening, and since endogenous ethylene is generated only after the fruits have dropped, this character might
have economic value if transferred to fresh-market cultivars
intended for storage and long-distance transport. It is too
early to speculate on potential utilizations of these discov-
eries, but clearly further research is justified on respira-
tion cycles of the wild species and of their hybrid derivatives x L. *esculentum* to determine the behavior patterns after
introgression into the genetic framework of the latter.

Self-incompatibility

Attempts have been made to transfer self-incompatibility to
L. esculentum from *L. chilense* (Martin, 1961) and from *L. hirsutum* (Martin, 1968). Thanks to the action of several major
dominant genes, *S* and other regulators, self-incompatible
segregants could be identified with relative ease in the early
backcrosses of *chilense* alleles, but a gradual weakening of
the reaction in later backcrosses indicated action of additional, minor genes. In the cross with *L. hirsutum*, although
a strong self-incompatibility reaction persisted, pollen-
specific properties were gradually lost with continued introgression. On the basis of these experiences, Martin concludes
that "the prospects of producing self-incompatible *L. esculentum* populations therefore appear to be small."

Despite Martin's (1961, 1968) discouraging experiences,
Denna (1971) proposed programs intended to introgress self-
incompatibility from wild sources in order to facilitate large
scale production of F_1 hybrid seed. In a system to utilize
self-incompatibility in this fashion, introgression from the
wild parent of genes regulating not only this character but
also certain floral characters must be accomplished in order
to attract pollinating insects and to make their visits effective for cross-pollination. In an investigation of the
interrelationships of such traits and self-incompatibility in
several tomato species, control of the critical traits, e.g.,
stigma exsertion and size of flower and inflorescence, was
found to be polygenic and inherited largely independently of
the self-incompatibility loci (Rick, 1982). An additional
consideration, i.e., the restoration of self-fertility and
guarantee of high fruitfulness in the F_1 hybrids, poses another difficulty that must be hurdled to implement such a
program. Considering the complexity of these problems, the
prospects at this time indeed do not look bright for exploiting self-incompatibility in this fashion.

Amplified Allozymic Variability

Allozymic markers are highly useful for various applications in plant breeding because they are 100% heritable,
generally lack epistatic interactions and detectable effects
on morphology or performance, and are expressed in very early
developmental stages. Yet allozymic variation essential for
such purposes is lacking in the germplasm of older European
and North American stocks of the cultivated tomato, presumably as a result of its history of domestication (Rick and
Fobes, 1975). An abundant supply of such variation can be

supplied, however, by controlled introgression from the related wild species. As an example, the exceedingly tight
linkage between *Aps-1* (acid phosphatase-1 locus) and *Mi*
(dominant gene coding for resistance to many biotypes of root-
knot nematode) is being widely exploited by scoring the *Aps-1*
segregation as a means of selecting more efficiently for resistance (Rick and Fobes, 1975; Medina-Filho, 1980), both
monogenic variants having been derived from an accession of *L.
peruvianum* (Gilbert and McGuire, 1956). In a similar situation, thanks to the linkage distance of about 1 cM between
Prx-2 (peroxidase-2 locus) and *ms-32* (a highly useful recessive male sterility), selecting the appropriate *Prx-2* phenotypes facilitates consecutive backcross transfers of *ms-32*
and production of nearly 100% male-sterile progenies (Tanksley and Rick, unpublished). In another application, advantage is taken of the simultaneous segregation of 12 isozymic
loci on nine chromosomes in the wide cross *L. esculentum* x *S.
pennellii* to select more effectively for a rapid reconstitu-
tion of the genotype of the recurrent parent in backcrosses to
the former species (Tanksley et al., 1981) and to map and analyze action of genes controlling various quantitative characters (Tanksley et al., 1982).

VI. DISCUSSION

    In contemplating this enumeration of genetic resources al-
ready revealed in the tomato species, one cannot fail to be
impressed by the great opportunities that exist for further
improvement of the cultivated tomato. The accomplishments to
date by tomato breeders are no less impressive. In the area
of disease control alone, resistance to at least 14 pathogens
has been uncovered in these sources and bred into existing
cultivars, many of which are now grown on a large scale. Also, as the examples of introduction of improved fruit quality
reveal, much progress has also been recorded in this area.
    In respect to this formidable record, the tomato may rank
at the top of the list of crop plants improved by introgression of genes from wild or exotic sources. It affords a sharp
contrast to the history of improvement of maize, in which crop
breeders have generally found the genetic resources needed for
their objectives within only a portion of the species, i.e.,
certain races native to North America. Reasons for the extreme contrast between these crop plants are not difficult to
find. In the first place, tomatoes are autogamous and maize,
allogamous, fostering the preservation of much genetic variation within populations. Secondly, maize has been bred

continuously for centuries within its native area. In contrast, most tomato improvement for the past four centuries
took place outside the native area in a tiny segment of the
total genetic reserves existing in the species. Migration of
the ancestral var. *cerasiforme* from the Andes through Central
America to Mexico, followed by introduction of the domesticates to Europe and later back to the New World must certainly have restricted the genetic base. Even without selection,
passage through these many bottlenecks would have led to depletion of genetic variability (Rick and Fobes, 1975). Other
crops of New World origin that followed similar routes and
probably suffered genetic losses are the peppers (*Capsicum*
spp.), squashes (*Cucurbita* spp.), bean species (*Phaseolus*
spp.), potato (*Solanum tuberosum*), peanut (*Arachis hypogea*),
and sweet potato (*Ipomea* spp.).

Novel Variation

At several places herein reference has been made to the
appearance in interspecific hybrid progenies of characteristics that were not detected in the parents. A classic example is the appearance of orange color in fruits encoded by *B*
from *L. hirsutum*, in which no phenotypic manifestation of
this color has yet been detected. This phenomenon can be explained satisfactorily by interaction between *B* and the background genotype of *L. esculentum*. Other examples of such
novel characters are:
1. *Ip*, the pigment intensifier derived from *L. chmielewskii* (Rick, 1974).
2. Male sterility derived from *L. parviflorum* (Kesicki,
    1980).
3. Calyx abnormalities accompanying *j-2* derived from *L.
    cheesmanii* (Rick, 1967).
Transgressive segregation, which is the manifestation of
novel variation in quantitative characters, is also commonly
encountered in the progenies of tomato interspecific hybrids.
Examples are:
1. Leaf water retention derived from *S. pennellii* (Yu,
    1972).
2. Soluble solids content of fruits derived from *L.
    chmielewskii* (Rick, 1974).
3. Ability to flower in short-day, low light intensity
    conditions derived from *L. hirsutum* (Maxon-Smith,
    1978).
This type of genetic behavior is so common in interspecific hybrids that it may be more in the nature of a rule than
an exception. The point will not be belabored here, but

attention must be called to the occurrence of this phenomenon, which can often add measurably to success in selecting quantitatively inherited characters. Without previous experience, novel traits by definition cannot be predicted. Their appearance being fairly frequent, it behooves the investigator to be constantly vigilant for them and to exploit them at appropriate opportunities. They manifest the need for studying segregating progenies, primarily from selfed derivatives, in order to evaluate fully the potential of wild or exotic accessions.

Genetic Resources in the Wild Species

The examples described here represent only a tiny fraction of the extent to which genetic reserves in the tomato species could be exploited. Most of the cited instances trace to past hybridizations dating back as far as five decades. Since that time the collections of tomato species have been augmented by a factor of many times. The more recent collections have, furthermore, been made and reproduced by improved procedures to maintain a maximum of existing genetic variability.

Experience has already revealed great opportunities in the utilization of wild germplasm. The tomato species are not homogeneous entities. To be sure, certain traits (like taxonomic key characters) are fixed in the species. Thus, leaf water retention is universal in *S. pennellii; B* is fixed throughout *L. hirsutum* and other greenfruited species; and a high level of insect resistance seems to be universal in *L. hirsutum*. In contrast, a vast number of heritable traits are variable. Recent studies on the nature of genetic variation in the tomato species have revealed unexpected situations in respect to the nature of mating systems, extent of variation, etc. The best understanding has been provided by investigations of allozyme variability, which indicate that significant variation exists between accessions of all wild species. Nearly absolute autogamy is encountered in *L. cheesmanii, L. parviflorum*, and marginal populations of *L. hirsutum, peruvianum*, and *pimpinellifolium*. Evaluation of these groups is simplified by the remarkable genetic uniformity within wild populations. At the other extreme, the self-incompatible types of *L. chilense, L. hirsutum, L. peruvianum*, and *S. pennellii* are subject to a bewildering extent of variability as observed within and between populations, and between races. Complete evaluation of the genetic resources in these materials requires very large samples. An intermediate level of variability is generally characteristic of the self-fertile, but facultatively outcrossing forms of *L. chmielewskii* and *L. pimpinellifolium*. The implications of these differences have

been considered elsewhere (Rick, 1976), and hence do not war-
rant elaboration here.

To exploit these resources in a thorough fashion would be
an enormous task. In these times, when fewer, rather than
more, resources are being devoted to germplasm utilization,
the prospects for any kind of justice to these untapped resources are bleak. In the chain of events that transpires
between discovery of a useful trait in the wild species and
introduction of the improved cultivar, a serious gap exists
in the initial steps of transferring wild traits to the genetic background of the cultivated species. This procedure,
commonly referred to as "prebreeding," usually consists of a
series of backcrosses to the recurrent, cultivated parent al-
ternated with selfed generations, in which desired combinations of parental characteristics are selected. It permits
evaluation of the introgressed traits and facilitates breeding improved cultivars. It also provides opportunity for detecting transgressive segregation, interspecific epistasis,
and other types of novel variation. More incentive and more
support are vitally needed to expand prebreeding activities
in the tomato species as well as in the wild relatives of
many crop species, in order to place advanced, improved lines
in the hands of crop breeders.

SUMMARY

The tomato (*Lycopersicon esculentum*) is a prime example of
a cultivated plant that benefits from experimental introgression of germplasm from wild and exotic taxa. For various
historical reasons the total variation existing in European
and North American stocks has been severely limited. Great
improvement has already been wrought via incorporation of improved fruit quality and of resistance to at least 14 diseases in the breeding of cultivars that are currently grown
on a large scale. Resistance to many other diseases and to
pernicious arthropods; tolerance of such environmental
stresses as temperature extremes, deficient and excessive
moisture, and salinity; improved fruit quality in respect to
pigment, sugar, acid, volatile, and nutritional content; and
an additional series of useful morphological and physiological characters have been identified in these sources. Much
research is being conducted on the transfer of these traits
to *L. esculentum* and on their expression in the wild sources
and introgressants. These endeavors are greatly expedited by
several factors. (1) Living material of a large array of accessions of each species is available. Assays of morphological and allozymic characters reveal extensive genetic

variation in most of the species. (2) *L. esculentum* can be
crossed with the eight other tomato species with varying
degrees of difficulty, and the resultant hybrids are
sufficiently fertile to yield the required progenies. (3)
Transgressive and other novel variation augment the range of
variability in the segregating progenies.

The purpose of this article is to review the utilization
of tomato species and exotic cultivars of *Lycopersicon esculentum* for improvement of the latter. Such areas previously
considered, e.g., stress tolerance (Rick, 1973), will be up-
dated rather than reviewed *in extenso*.

It is fortunate for tomato workers that a vast reserve of
genetic resources exists in related wild species. The cultivated species (*L. esculentum*) itself is represented by the
feral and wild var. *cerasiforme*. Eight other species known
only in the wild state, serve as additional germplasm resources. Some of these species can be readily mated with *L.
esculentum*; for other crosses, special aids, such as embryo
culture, are required. The F_1 hybrids have sufficient fertility to generate F_2 or Bc progeny; although the fertility of
some combinations is low, the potential seed yield of the
parental species is so large that it is usually not difficult
to obtain sufficient offspring for breeding or genetic purposes. Fertility and vigor of the hybrid progenies are also
generally adequate for those purposes. This work is also expedited by the homosequentiality of chromosomes in all nine
species.

A review of this subject is fraught with several difficulties. One of them concerns the problem of tracing the source
of various improved traits incorporated in new cultivars.
Pedigrees may be lacking, or, if available, often lead to
parents of unknown origin. Further, even if complete, multi-
parental pedigrees may not reveal which of several parents is
the source of the trait in question.

ACKNOWLEDGMENTS

Editorial assistance by Dora G. Hunt and useful comments
by Dr. M. Allen Stevens are gratefully acknowledged.

REFERENCES

Acosta, J. C., Gilbert, J. C., and Quinon, V. L. (1964). *Proc. Amer. Soc. Hort. Sci. 84*, 455.

Alexander, L. J. (1963). *Phytopathology 53*, 869.

Anon. (1971). Glasshouse Crops Research Institute 1971-- Annual Report.

Andrus, C. F., and Reynard, G. B. (1945). *Phytopathology 35*, 16.

Andrus, C. F., Reynard, G. B., Jorgensen, H., and Eades, J. (1942). *J. Agric. Res. 65*, 339.

Barksdale, T. H., and Stoner, A. K. (1978). *Plant Dis. Reptr. 62*, 844.

Bliss, F. A., Onesirosou, P. T., and Arny, D. C. (1973). *Phytopathology 63*, 837.

Bohn, G. W., and Tucker, C. M. (1940). *Mis. Agric. Expt. Sta. Res. Bul. 311*, 1.

Crill, P., Bryan, H. H., Everett, P. H., Bartz, J. A., Jones, J. P., and Matthews, R. F. (1977). *Florida Agric. Expt. Sta. Circular S-248*.

Dehan, K., and Tal, M. (1978). *Irrig. Sci. 1*, 71.

Denna, D. W. (1971). *Euphytica 20*, 542.

Dinar, M., and Stevens, M. A. (1981). *J. Amer. Soc. Hort. Sci. 106*, 415.

El Ahmadi, A. B., and Stevens, M. A. (1979a). *J. Amer. Soc. Hort. Sci. 104*, 686.

El Ahmadi, A. B., and Stevens, M. A. (1979b). *J. Amer. Soc. Hort. Sci. 104*, 691.

Epstein, E., Norlyn, J. D., Rush, D. W., Kingsbury, R. W., Kelley, D. B., Cunningham, G. A., and Wrona, A. F. (1980). *Science 210*, 399.

Fery, R. L., and Cuthbert, F. P. (1975). *J. Amer. Soc. Hort. Sci. 100*, 276.

Finlay, K. W. (1953). *Austral. J. Biol. Sci. 6*, 153.

Frederickson, D. L., and Epstein, E. (1975). *Plant Phys. 56*, 4.

Gallegly, M. E., and Marvel, M. E. (1955). *Phytopathology 45*, 103.

Gentile, A. G., and Stoner, A. K. (1968a). *J. Econ. Entomol. 61*, 1152.

Gentile, A. G., and Stoner, A. K. (1968b). *J. Econ. Entomol. 61*, 1347.

Gentile, A. G., Webb, R. E., and Stoner, A. K. (1968). *J. Econ. Entomol. 61*, 1355.

Gentile, A. G., Webb, R. E., and Stoner, A. K. (1969). *J. Econ. Entomol. 62*, 834.

Gilardón, E. M., and Benavent, J. M. (1981). *Soc. Argentina Olericult. (Salta)*, p. 12 (abstr.).

Gilbert, J. C., and McGuire, D. C. (1956). *Proc. Am. Soc. Hort. Sci. 68*, 437.

Grimbly, P. E. (1981). Abstracts, Eucarpia Tomato Working Group, Avignon, p. 9.

Grummet, R., Fobes, J. F., and Herner, R. C. (1981). *Plant Phys.* (in press).

Hassan, A. A., Mazyad, H. M., Moustafa, S. E., and Nakhla, M. K. (1981). *Egypt. J. Hort.* (in press).

Hendrix, J. W., and Frazier, W. A. (1949). *Hawaii Agric. Expt. Sta. Bull. 8*, 1.

Hewitt, J. D., Dinar, M., and Stevens, M. A. (1982). *J. Amer. Soc. Hort. Sci.* (submitted).

Hewitt, J. D., and Stevens, M. A. (1981). *J. Amer. Soc. Hort. Sci.* (in press).

Hódosy, S. (1975). *Kertgazdaság 7(2)*, 51.

Hogenboom, N. G. (1970). *Euphytica 19*, 413.

Hogenboom, N. G., de Ponti, O. M. B., and Pet, E. (1974). Eucarpia Meeting, Bari, Italy, p. 138.

Jacquemond, M., and Laterrot, H. (1981). Abstracts, Eucarpia Tomato Working Group, Avignon, p. 12b.

de Jong, J., and Honma, S. (1976). *Euphytica 25*, 405.

Juvik, J. A. (1980). Ph.D. Thesis, University of California, Davis.

Katz, A., and Tal, M. (1980). *Z. Pflanzenphysiol. 98*, 429.

Kennedy, G. G., and Henderson, W. R. (1978). *J. Amer. Soc. Hort. Sci. 103*, 334.

Kennedy, G. G., and Yamamoto, R. T. (1979). *Ent. Exp. Appl. 26*, 121.

Kerr, E. A., and Bailey, D. L. (1964). *Can. J. Bot. 42*, 1541.

Kesicki, E. (1980). *Actae Horticult. 100*, 365.

Koriyama, T., and Kuriyasu, K. (1974). *Bul. Veg. Otu. Crops Res. Sta.* (Hokoku, Japan) *A(1)*, 93.

Laterrot, H., and Rat, B. (1981). Abstracts, Eucarpia Tomato Working Group, Avignon, p. 19b.

Laterrot, H., Brant, R., and Daunay, M. C. (1978). *Ann. d'Amel. Plantes 28*, 579.

Lincoln, R. E., and Porter, J. W. (1950). *Genetics 35*, 206.

Makmur, A., Gerloff, G. C., and Gabelman, W. H. (1978). *J. Amer. Soc. Hort. Sci. 103*, 545.

Martin, F. W. (1961). *Genetics 46*, 1443.

Martin, F. W. (1968). *Genetics 60*, 101.

Martin, M. W. (1970). *Euphytica 19*, 243.

Mathew, J., and Patel, P. N. (1975). *Curr. Sci. 44*, 867.

Maxon-Smith, J. W. (1978). *In* "Interspecific Hybridization in Plant Breeding" (E. Sanchez-Monge, F. Garcia-Olmedo, eds.), p. 119. Proc. 8th Congr. Eucarpia, Madrid.

Medina-Filho, H. (1980). Ph.D. Thesis, University of California, Davis.

O'Sullivan, J., Gabelman, W. H., and Gerloff, G. C.   (1974).
   J. Amer. Soc. Hort. Sci. 99, 543.
Patterson, B. C., Paull, R., and Smillie, R. M.   (1978).
   Austral. J. Plant Physiol. 5, 609.
Patterson, C. G., Knavel, D. E., Kemp, T. R., and Rodriguez,
   J. G.   (1975).   *Envt. Ent. 4*, 670.
Paull, R., Patterson, B., and Payue, L.   (1979).   *Plant Phys.
   63* (5 Suppl.), 102 (abstract).
Pilowsky, M., and Cohen, S.   (1974)   *Phytopathology 64*, 632.
Porte, W. S., and Wellman, H. B.   (1941).   *USDA Circ. 611*, 1.
Rattau, R. S., and Saini, S. S.   (1979).   *Euphytica 28*, 315.
Rebigan, J. B.,Villareal, R. L., and Lai, S. H.   (1977).
   Philipp. J. Crop Sci. 2, 221.
Richards, M. A., Phills, B. R., and Hill, W.   (1979).
   HortScience 14, 121.
Rick, C. M.   (1967).   *Econ. Bot. 21*, 171.
Rick, C. M.   (1973).   *In* "Genes, Enzymes, and Populations"
   (A. Srb, ed.), p. 255.   Plenum, New York.
Rick, C. M.   (1974).   *Hilgardia 42*, 493.
Rick, C. M.   (1976).   *Genet. Agraria 30*, 249.
Rick, C. M.   (1982).   *Biol. Zbl.* (in press).
Rick, C. M., and Fobes, J. F.   (1974).   *Rept. Tomato Genet.
   Coop. 24*, 25.
Rick, C. M., and Fobes, J. F.   (1975).   *Bul. Torrey Bot. Club
   102*, 376.
Robinson, R. W., and Kowalewski, E.   (1974).   *Proc. 19th Int.
   Hort. Congr. (Warsaw) 1B*, 716.
Rosen, A., and Tal, M.   (1981).   *Z. Pflanzenphysiol.* (in
   press).
Rush, D. W., and Epstein, E.   (1976).   *Plant Phys. 57*, 162.
Rush, D. W., and Epstein, E.   (1980).   *Plant Phys. 63* (6
   Suppl.), 83.
Saccardo, F., and Ancora, G.   (1981).   Abstracts, Eucarpia
   Tomato Working Group, Avignon, p. 25.
Schaible, L., Cannon, O. S., and Waddoups, V.   (1951).   *Phy-
   topathology 41*, 986.
Schultz, J. H.   (1953).   *N. D. A.E.S. Bimonthly Bul. 15(3)*,
   114.
Schuster, D. J.   (1977).   *J. Econ. Entomol. 70*, 434.
Schuster, D. J., Waddill, V. H., Augustine, J., and Volin, R.
   B.   (1979).   *J. Amer. Soc. Hort. Sci. 104*, 170.
Sinden, S. L., Schalk, J. M., and Stoner, A. K.   (1978).   *J.
   Amer. Soc. Hort. Sci. 103*, 596.
van Steekelenburg, N. A. M.   (1981).   Abstracts, Eucarpia
   Tomato Working Group, Avignon, p. 28.
Steele, A. E.   (1977).   *J. Nematology 9*, 285.
Stevens, M. A.   (1970a).   *HortScience 5*, 95.

Stevens, M. A.   (1970b).   *J. Amer. Soc. Hort. Sci. 95*, 9.

Stevens, M. A.   (1970c).   *J. Amer. Soc. Hort. Sci. 95*, 461.

Stevens, M. A.   (1972).   *J. Amer. Soc. Hort. Sci. 97*, 655.

Stevens, M. A.   (1974).   *In* "Nutritional Qualities of Fresh
    Fruits and Vegetables" (P. L. White, ed.), p. 87. Futura,
    New York.

Stevens, M. A., Kader, A. A., Albright-Holton, M., and Algazi,
    M.   (1977).   *J. Amer. Soc. Hort. Sci. 102*, 680.

Stevens, M. A., and Rudich, J.   (1978).   *HortScience 13*, 673.

Stevenson, W. R., Evans, G. E., and Barksdale, T. H.   (1978).
    Plant Dis. Rept. 62, 937.

Tal, M.   (1971).   *Austral. J. Agric. Res. 22*, 631.

Tal, M., and Katz, A.   (1980).   *Z. Pflanzenphysiol. 98*, 283.

Tal, M., Heikin, H., and Dehan, K.   (1978).   *Z. Pflanzenphys-
    iol. 86*, 231.

Tal, M., Katz, A., Heikin, H., and Dehan, K.   (1979).   *New
    Phytol. 82*, 349.

Tanksley, S. D., Medina-Filho, H., and Rick, C. M.   (1981).
    Theor. Appl. Genet. (in press).

Tanksley, S. D., Medina-Filho, H., and Rick, C. M.   (1982).
    Heredity (submitted).

Tomes, M. L.   (1963).   *Bot. Gaz. 124*, 180.

Tomes, M. L., and Quackenbush, F. W.   (1958).   *Econ. Bot. 12*,
    256.

Vallejos, C. E.   (1979).   *In* "Low Temperature Stress in Crop
    Plants" (J. Lyons, ed.), p. 473. Academic Press, New York.

Webb, R. E., Stoner, A. K., and Gentile, A. G.   (1971).   *J.
    Amer. Soc. Hort. Sci. 96*, 65.

Williams, W. G., Kennedy, G. G., Yamamoto, R. T., Thacker, J.
    D., and Bordner, J.   (1980).   *Science 207*, 888.

Yeager, A. F., and Purinton, H. J.   (1946).   *Proc. Amer. Soc.
    Hort. Sci. 48*, 403.

Yu, A. T. T.   (1972).   Ph.D. Thesis, University of California,
    Davis.

Zamir, D., Tanksley, S. D., and Jones, R. A.   (1981).   *Theor.
    Appl. Genet. 59*, 235.

CHAPTER 2

EXOTIC GERMPLASM
IN CEREAL CROP IMPROVEMENT

William L. Brown

Pioneer Hi-Bred International, Inc.
Des Moines, Iowa

Interest in the use of exotic germplasm for the improvement of crop species, including cereals, increased dramati-
cally following the Southern Corn Leaf Blight epidemic on
maize (*Zea mays*) in the USA in 1969-70. This epidemic is
estimated to have caused a reduction in U.S. maize yield of
15% nationwide and as much as 50% in some southern states
(Tatum, 1971).
    Although neither the first nor the most severe epidemic
to strike an important economic species, Southern Corn Leaf
Blight aroused sufficient fear to cause those concerned about
adequate supplies of food and fiber to look seriously at the
amount of genetic diversity present in the widely used cultivars of important crops. A comprehensive study by a special
committee of the National Academy of Sciences of the United
States found that many U.S. crops did indeed rest on a relatively narrow genetic base, and consequently, were genetically
vulnerable (National Academy of Sciences, 1972).
    It should be noted that while the term "genetic vulnera-
bility" has come into common usage in the past decade, the
subject is not new. The importance of genetic diversity and
the potential consequences associated with narrowing of the
germplasm base has long been recognized by most plant breeders. They do not always agree that the genetic base of the
species with which they work is as narrow as some non-breeders
believe, yet they are aware of the problem. Any sound breeding program must take into account the response of new selections to hazards of the environment before such selections

are released for general use. A primary purpose of testing
and screening is to evaluate the vulnerability of potentially
new cultivars to disease and insect pests and to other
stresses imposed by the environment. The estimation of cultivar x year and cultivar x location interactions is one way of
predicting genetic vulnerability of genotypes. So, among
plant breeders, the problem of genetic vulnerability is not
one of recent origin. Only the increased and widespread interest in the problem is new. And with this growing interest
there is an increasing awareness of the possible role of
exotic germplasm in reducing genetic vulnerability. If genetic vulnerability is due to a decrease in genetic variability,
the most obvious way to increase variability is by introducing germplasm from distantly related species, exotic germplasm, into breeding populations of the cultivated crop. For
discussion, I prefer to define exotic germplasm as Hallauer
and Miranda (1981) did, i.e., to "include all germplasm that
does not have immediate usefulness without selection for
adaptation to a given area."

Before considering the use of exotic germplasm for the improvement of specific cereals, comment needs to be made on
how exotic strains might best be chosen and the difficulty
encountered in incorporating such lines into adapted populations. Clearly there is no reason to use exotic germplasm
in a breeding program simply because it is exotic. Such
materials are oftentimes chosen because of their known resistance to specific diseases or insects or as known sources of
other heritable traits of agronomic value. If the breeder is
searching for improved overall performance, a knowledge of
the evolutionary development of the wild relatives, land
races, and cultivars that comprise the species under consideration can be helpful in determining the types of exotic germplasm that are likely to be most useful.

Most crop species have evolved through introgression--
introgression between species, between distinct races within
species, or from primitive or weedy forms. An understanding
of these evolutionary processes and the geographic areas in
which they occurred can have important implications for
breeding. If, for instance, the dominant elements in the
evolutionary development of a highly productive race are
known, this should tell something about where to look for
maximum heterosis within that race. If primitive or weedy
forms played a role in the evolution of a species, one might
expect those same weedy forms to be useful in the further improvement of the species. For example, the chances of improving bread wheat might be enhanced if the putative ancestors
of *Triticum aestivum* were used more extensively in wheat
breeding programs (Feldman and Sears, 1981). In maize, additional introgression of teosinte (*Zea mexicana*), either

directly or through teosinte-containing maize cultivars,
might provide further improvement of those cultivars that
contain little teosinte in their genetic makeup.

The point is that understanding the evolutionary history
of a crop species should tell a breeder which sources of
germplasm have been most influential in making the species
what it is. It should tell how to look for such sources of
germplasm and where they are most likely to be found. For
example, there is evidence that among the recognized maize
land races from Mexico, the Caribbean, North America, and
parts of South America, those that are most productive share
a common evolutionary pattern of development (Brown, 1975).
The most important element of that pattern is a long history
of interracial introgression involving numerous races. Moreover, in Mexico, most if not all of the highest yielding races
have teosinte ancestries. Not all maize races have the complicated ancestry that seems to characterize the most productive ones. Several such races evaluated in maize improvement
programs, even though they may contain useful genes or combine well with unrelated races, do not, to my knowledge, possess above average yielding ability.

Thus, for species for which the evolutionary development
of land races is known and for which evaluation data are
inadequate or non-existent, exotic races that arose through
introgression would seem to be better choices than those of a
more simple ancestry. The introduction of such strains into
breeding populations should both increase genetic variability
and enhance the potential yielding capacity of the derived
populations. Genetic variability of a population can be increased easily by introgressing into it germplasm from any
unrelated source. However, increasing genetic variation by
extending the yield range at the low end of the distribution
is of no interest to the breeder. So, if exotic germplasm is
used, the exotic strains should be chosen with attention equal
to that given the elite, adapted parents.

Even if the exotic germplasm chosen for introgression is
elite, the problem of introgressing such germplasm into
adapted genotypes is considerable. Most land races and all
wild relatives of cultivated crops are poor agronomic types
that contain many undesirable alleles that must be eliminated
from progeny prior to commercial use. The frequent presence
of tight linkages between desirable and undesirable alleles
can complicate the selection process. Attempts to use lines
adapted to the tropics in the temperate zone usually result
in problems of day-length response which effectively mask the
expression of other traits of interest. This essentially
prevents the evaluation of exotic materials in an environment
to which they are unadapted. Tight linkages of loci probably

cause the inordinately large number of generations required
to obtain normal recombination between genes from exotic and
adapted strains relative to the number required for progeny of
matings between adapted genotypes (Lonnquist, 1974). Therefore, the time required to obtain usable material from exotic
or exotic x adapted populations of segregates is greater than
the time required for adapted populations. Partly for these
reasons, exotic germplasm has been used only to a limited extent.

I. MAIZE (*Zea mays*)

Examples of the successful use of exotic maize germplasm
in the improvement of the commercial crop are not numerous
despite the fact that most maize breeders have at sometime
mated elite lines and cultivars, with exotic strains. Probably many reasons account for the limited success with maize
introgression research, but at least two are obvious. First,
some exotic maize strains used have been so poor in agronomic
traits and yielding ability that they could not possibly contribute positively to elite, adapted genotypes. Second, the
breeding procedures generally employed in exotic germplasm
introgression probably would preclude the generation of segregants of commercial merit. Following the mating of exotic x
elite genotypes the breeder usually selfs the F_1 and succeed-
ing generations (a type of pedigree selection procedure) because this method has been so effective when working with
progeny from matings of elite, adapted parents. This procedure invariably produces disappointing results because
several generations of intermating to break up linkages were
not included prior to selfing. Segregants obtained with the
pedigree selection procedure usually differ little from the
parental strains mated.
   Despite the many disappointing experiences, both reported
and unreported, in using exotic germplasm for maize improvement, several examples suggest that exotic lines can be used
effectively to increase genetic variability and as sources
for improving agronomic traits, including yield. Griffing
and Lindstrom (1954) compared the performance of hybrids
derived by mating lines within and between three sources (a)
Corn Belt, (b) Brazilian, and (c) Corn Belt with 25% and 50%
Mexican germplasm. Hybrids containing exotic germplasm,
either wholly or in part, were higher yielding than those
containing only Corn Belt germplasm.
   Goodman (1965) reported genetic variances for two populations developed by the author. These populations were developed in such a way that direct comparisons could be made of

the effect of the introgression of Caribbean into Corn Belt germplasm. Ten elite lines from the Corn Belt were inter-mated to form a Corn Belt population. The same lines were mated with a group of West Indian races to develop a second breeding pool (West Indian population). The two populations were grown and mass selected simultaneously in Iowa, USA, for ten generations. Particular attention was given to selecting those phenotypes which on the basis of morphology appeared to be recombinant genotypes. Goodman's (1965) study, conducted in Iowa and North Carolina, showed that both genetic variabil-ity and expected gain from selection were greater in the West Indian than in the Corn Belt population. In North Carolina, mean yields of the West Indian population were greater than those of the Corn Belt population, whereas the reverse was true in Iowa. However, and perhaps more importantly, the highest yields for full-sib families from the West Indian population were greater than the highest ones from the Corn Belt population, both in Iowa and North Carolina. These re-sults suggest, therefore, that the introduction of exotic germplasm into elite, well-adapted Corn Belt genotypes gave both increased genetic variability and greater yielding ability as well.

On the basis of grain yields from two diallel series of matings involving Corn Belt and semi-exotic populations (including the West Indian population), Eberhart (1971) con-cluded that the better semi-exotic populations were promising sources of useful germplasm for temperate zone maize improve-ment programs.

Nelson (1972) developed a number of inbred maize lines of commercial value from mating of adapted genotypes with vari-ous exotic races including Tuxapeno, Cuban Flint, Coastal Tropical Flint, and other races or cultivars from Colombia, Peru, and Brazil. These derived lines have provided wide adaptation to southern USA, excellent combining ability, and drought and disease tolerance. Nelson (1972) emphasized the importance in his breeding procedure of sibbing among progeny of the original matings for seven or more generations before selfing was initiated.

Two other examples of the successful use of exotic germ-plasm in the improvement of maize in USA Corn Belt programs are reported by Seifert (1981). The first involved the race Cuban Flint. In the mid 1950's Cuban Flint was crossed with Conn. 103, a widely used U.S. inbred developed from the open pollinated cultivar, 'Lancaster Surecrop'. Cuban Flint was used to introduce ear worm resistance into locally adapted maize lines. By using a procedure of alternate sibbing and selfing in the progeny of the mating, an inbred line was developed which had commercial value and was used as one parent of a successful double-cross hybrid adapted to southern

USA. Subsequently, this line was mated with another from the central Corn Belt, and from the progeny, a second cycle line was selected which, according to Seifert, is probably the most widely adapted line in the group used by a large commercial company. The line has good insect tolerance and acceptable yields. Interestingly, this second cycle inbred line contains, on a theoretical basis, only one-quarter exotic germplasm: This suggests that the introduction of only small increments of exotic germplasm into adapted genotypes may be the most effective way of utilizing such races.

A second example cited by Seifert (1981) may indicate the potential of related species for maize improvement. The recently described teosinte, *Zea diploperennis* (Iltis, 1979), seems to be highly tolerant or resistant to a number of important virus diseases that attack maize (maize chlorotic dwarf, maize dwarf mosaic, etc.). Seemingly, genes for virus resistance from *Z. diploperennis* can be transferred to maize without carrying along the many undesirable traits present in teosinte.

Earlier, Reeves (1950) used annual teosinte as a source of genes for maize improvement. He crossed Texas inbred 127C with teosinte, backcrossed to the maize, and selfed the progeny to obtain lines with varying amounts of teosinte germplasm. Certain derived lines, when evaluated in topcrosses, yielded 20% more than 127C hybrids and were more resistant to drought. However, there is no record of these lines having been used in commercial maize hybrids.

Despite the need to increase genetic diversity in maize breeding populations and reports of positive results from using exotics, there is limited interest among maize breeders in using exotic germplasm. Several factors cause this reticence, but the most practical reason is the consistent improvement being made via the use of adapted, non-exotic germplasm sources.

Russell (1974) and Duvick (1977) have shown that maize breeders of the USA, using adapted germplasm almost exclusively, have made significant and consistent genetic improvement in the yields of maize hybrids over the past 50 years. Genetic gains in yield over that period have averaged 1 q/ha per year. In addition, the rate of yield improvement does not appear to be approaching a plateau. So, as long as this rate of genetic gain can be maintained, the maize breeder, for good reasons, will concentrate breeding efforts on those adapted sources of germplasm of proven effectiveness. Undoubtedly, this will be the most efficient approach in the short run, but, long term progress will be greater, in my estimation, if the bases of the maize breeding pools are broadened by introgression from a larger sample of genetic sources.

The sections on wheat, sorghum, oats, and rice will not be comprehensive reviews of the use of exotic germplasm for improving these crops. Much data have been published on the value of exotic germplasm for improving these species, so this review will highlight a few examples of significant improvement in yield or other traits that have resulted from introgressing exotic or semi-exotic germplasms into breeding populations.

II. WHEAT (*Triticum aestivum*)

Bread wheat, *Triticum aestivum*, which is cultivated worldwide and is a staple food for much of the world's human population, cannot be traced to a single wild progenitor but is, instead, the genetic product of combining exotic germplasm from three widely different, but related, species (Riley, 1965).

Much effort in modern wheat breeding has been devoted to searching for new sources of disease resistance, and the sources of resistance have included both exotic and non-exotic collections. McFadden (1930), using conventional breeding techniques, transferred stem-rust resistance crossed from Yaroslov emmer, *T. dicoccum*, to Marquis, a cultivar of *T. vulgare*. The mating was made in 1916, and, following eight generations of selection, the line H49-24, a high yielding, stem-rust resistant segregant, was named Hope and released.

Feldman and Sears (1981) suggest that genetic improvement for yield of bread wheat has stagnated because (a) the germplasm pool of cultivated wheats has already been fully exploited with respect to yield improvement and (b) drastic reduction has occurred in the genetic variability for other traits of cultivated wheats associated with productivity. They proceed to build a strong case for the use of wild related species as a source for restoring genetic variation and enriching the gene pool of cultivated wheats.

Greater use of wild relatives of wheat as sources of useful genes certainly should be encouraged, but, in my estimation, it probably is incorrect to assume that the breeding pool of cultivated wheat has been fully exploited. Greater progress in wheat improvement has been made via breeding in the past 15 years than was made in the previous half century. It is my impression that breeding populations of cultivated wheats encompass much greater genetic diversity today than they did 10, 20, or 30 years ago. Wheat breeders long ago made use of the wheat world collection as a source of genes for disease resistance, but in recent years only, breeders

have begun to use nonadapted and exotic lines to improve the
yielding capacity of bread wheats. In my opinion, this
change in attitude has been due primarily to the worldwide
influence of the CIMMYT (Centro Internacional de Mejoramiento
de Maiz y Trigo) wheat improvement program. The CIMMYT program, centered in Mexico but making use of the world for
testing purposes, has constituted breeding populations of
wheat that include lines from anywhere around the globe providing they appeared to have merit in geographic areas to
which they were adapted. Breeders working this program have
no hesitancy to intermate classes of wheat, e.g., the cooper-
ative project with Oregon State University to make spring x
winter matings. Many new gene combinations have emerged
from this massive program of deliberate introgression among
wheat genotypes. The use of two generations per year in two
distinct environments in Mexico, which was designed primarily
to facilitate selection for resistance to pathogens, also
provided, more or less by accident, the opportunity to select
for photoperiod insensitivity. Finally, the introduction of
dwarfing genes from Norin 10, whose usefulness in the USA
was first demonstrated by Vogel et al. (1956, 1963), permitted the selection of semidwarf genotypes capable of with-
standing high rates of nitrogen fertilization without lodg-
ing. So, from this program came a level of useful genetic
variability far greater than that previously available. Not
only were the CIMMYT wheats an important element in the
"Green Revolution," but because of the unselfish sharing of
the CIMMYT germplasm, wheat breeders around the world have
access to these new breeding materials. Depending on the
geographic location of the individual breeder, the CIMMYT
germplasm may be exotic, semi-exotic, or completely adapted.
But whatever the classification, the products from this program have added a new dimension to wheat breeding in many
parts of the world. This is an excellent example of the
opportunities for increasing genetic diversity by combining
germplasm from elite cultivars from widely separated locations. A myriad of new recombinant genotypes may be expected
to emerge from such introgression. While such a system may
not substitute for the use of germplasm from wild relatives,
it certainly is a simpler and less time-consuming way of
enriching gene pools. And, it should have application to
any species with a wide geographic distribution.

III.   SORGHUM (*Sorghum bicolor*)

Sorghum was probably first domesticated in Africa from
where it spread to India, China, the Middle East, and Europe.
It reached the Caribbean during the slave trade and from
there two introductions, Kafir and Milo, were later brought
to the USA. Kafir and Milo, plus limited amounts of Hegari
and Feterita, comprise the bulk of germplasm found in hybrid
grain sorghum today.

The first introductions of Kafir and Milo from Africa were
tall, late, tropical, or semi-tropical cultivars. As long as
sorghum was harvested by hand and grown in the southern part
of the USA these traits did not preclude the use of the introduced cultivars. However, with the introduction of the grain
combine in the 1930's, growing of sorghum expanded to areas
with shorter growing seasons where early and short cultivars
were needed. About this time, Quinby and Karper (1945), from
studies on the inheritance of duration of growth in sorghum,
demonstrated that duration of growth in Milo was controlled
by genes at four independently inherited loci, MA, MA2, MA3,
and MA4. By manipulating dominant and recessive alleles at
three of these four maturity loci, the number of days to
flower can be adjusted from as low as 48 days to as high as
90 days. The latest cultivars are dominant at all four loci
and the earliest are recessive at the first three. Intermediate forms usually are recessive at two of the first three
loci. All sorghums known are dominant at locus four (Quinby,
1974).

Similarly, height of sorghum is controlled by four non-
linked brachytic dwarfing genes. Tallness is dominant or
partially dominant to shortness. The tall or 0-dwarf sorghums reach a height of 3 to 4 meters whereas 4-dwarf plants
are as short as 1 meter or less. Thus, by manipulating these
genes to give 1-dwarf, 2-dwarf, 3-dwarf, or 4-dwarf genotypes,
height of sorghum cultivars can be adjusted to meet any
reasonable need for which the crop is used (Quinby, 1974).
Three-dwarf sorghums, which are well adapted to combine har-
vesting, are used most frequently in agriculturally developed
countries.

This understanding of the inheritance of height and growth
duration in sorghum has greatly facilitated the use of exotic
germplasm for sorghum improvement in recent years. Conversion programs carried out by the U.S. Department of Agriculture and individual breeders, have converted many late, tall,
tropical cultivars to early, 3-dwarf genotypes adapted to
temperate zone environments. These introductions have resulted in the use of exotic germplasm in sorghum on a scale

that is probably unequalled in any other cereal species.
These tropical sources of germplasm have provided a level of
resistance not previously available to a wide spectrum of
sorghum diseases. Also, they have contributed genes for
grain quality not present in adapted cultivars, but more importantly and perhaps unexpectedly, these tropical sorghum
introductions are contributing to improved grain yield.
Clearly, the programs for converting tropical sorghums have
greatly enriched the breeding pools of sorghum cultivars of
the temperate zones.

IV. RICE (*Oryza sativa*)

   Much of what is presented about rice herein comes from
suggestions received from Dr. T. T. Chang[1] and Dr. Howard L.
Carnahan[2].
   The natural variation in the cultivated gene pools of rice
apparently has been adequate to permit significant genetic
improvement for yield and pest resistance in this crop. Consequently, breeders have not needed to use wild species or
other exotic forms. Nonetheless, there are a few examples in
which related species or exotic cultivars have been the source
of useful genes. In California, USA, Carnahan (1981) is using
Oryza rufipogon as a source of resistance to stem rot (caused
by *Sclerotium oryzeae*). The cultivar M9, a semi-dwarf genotype, contains dwarfing genes extracted from the tropical
variety IR-8, an IRRI (International Rice Research Institute)
cultivar completely unadapted to California conditions. M9
was grown on approximately 50% of the California rice acreage
in 1979 (Carnahan, 1981).
   Carnahan (1981) also reports the successful use of *indica*
cultivars as sources of the long grain characteristic in the
development of cultivars with this trait for use in California. L-201, the first long grain rice released for production
in California, is derived from crosses of locally adapted
cultivars x *indica* cultivars.
   Ting (1933) described the development of the cultivar
Yatsen-1 from a cross of *O. sativa* x *O. sativa* var. *fatua*.

[1]*T. T. Chang, International Rice Research Institute, Los
Baños, Laguna, The Philippines.*

[2]*H. L. Carnahan, Rice Experiment Station, P. O. Box 306,
Biggs, California, USA.*

Yatsen-1 has good tillering ability and tolerance to cool
temperatures and acid soils. According to Chang (1981),
Yatsen-1 did not attain wide commercial use but it has been
an excellent source of genes for rice programs in China.
From a population of *O. nivara* collected in U.P. state, India,
a few plants were found to be resistant to the grassy stunt
virus. This population was the only one among 100 collections
of *O. nivara* that contained resistant seedlings to this disease. This source of grassy stunt resistance has been
transferred in at least eight IR cultivars from the IRRI rice
breeding program (Chang, et al., 1975).

A male sterile wild rice, *O. sativa f. spontanea*, from
South China, has been the primary source of cytoplasmic male
sterility of many of the hybrid rices grown in China (Chang,
1979). Hybrid rice cultivars that utilize this source of
male sterility were grown on about 5 million ha in each year
during the period 1978-80.

Also in China, a semi-dwarf, high yielding cultivar Ya-
nung-zoo was developed from a cross of Nung-keng 6 x a white-
awned, wild rice from Hainan Island. This cultivar is reported to produce yields as high as 6 t/ha in China (Kwangtung Agric. and Forestry College, 1975). Thus while the
variability in the cultivated rice pool may be adequate to
permit dramatic improvement through breeding (the IRRI program being a good example), the effective use of wild and
semi-wild relatives is noteworthy.

V.   OATS (*Avena sativa*)

The evolution of cultivated hexaploid oats is still poorly
understood (Holden, 1976), a fact that may explain why oat
breeders have made little use of the various diploid and
tetraploid cultivated types. Generally, oat breeding, both
in Europe and North America, has relied almost exclusively on
selection within the progeny from matings of pure-line cultivars. The lack of significant genetic variability within and
between these cultivars has restricted progress in oat improvement although it has permitted the development of improved straw strength and better resistance to fungal pathogens.

Beginning in the 1920's and 1930's, *A. sativa* was crossed
with *A. byzantina*, a close relative, to transfer disease
resistance from *A. byzantina* to *A. sativa*. Yields of selected cultivars from this mating, when compared in common experiments, showed increases of 9% to 14% over pure-line cultivars
(Browning et al., 1964; Langer et al., 1978). Frey (1981)

credits the yield improvement to increased genetic diversity
resulting from introgression of the two species. Langer et
al. (1978) further point out that since the cross of *A. sati-*
va x *A. byzantina*, forty years have elapsed without any significant additional progress in yield improvement of hexaploid oats in the midwestern USA.

The next advance in oat improvement was of a magnitude
that must be considered a significant breakthrough and is
an outstanding example of the effective use of exotic germplasm.

In the 1950's *A. sativa* was crossed with the wild species
A. sterilis, thought to be the progenitor of *A. sativa*.   *A.*
sterilis is also a hexaploid that crosses readily with *A.*
sativa to produce fertile hybrids.   This cross was made to
transfer *A. sativa* genes for resistance to race 264 of crown
rust (*Puccinia coronata* var. *avenae*).   At that time no known
source of resistance to race 264 was available among cultivated oat collections.   Following a number of backcrosses,
lines were selected and tested for yield.   Somewhat surprisingly, many crown-rust resistant lines developed in this way
were superior in yield to modern cultivars of similar maturity.   Eleven such lines selected from the Bc_3 to Bc_5 generations had a yield improvement of 20% or more.   When tested
for the 4-year period 1972-76, all exceeded the yield of
Clintford, a commercial cultivar.   Using Clintford as 100%,
the mean yields of the selected lines over the 4-year period
ranged from 103% to 129%.   Additionally, yield gains were
achieved without sacrificing other important agronomic characters (Frey, 1981).   Moreover, there also appears the possibility for increasing seed protein percentages in lines
derived from the *A. sativa* x *A. sterilis* matings.   It is
interesting to note that both *A. byzantina* and *A. sterilis*
were introduced into the breeding pools for the purpose of
adding disease resistance, yet significant yield gains were
achieved also.   This, it seems, should carry an important
message for breeders of oats and perhaps other crops as well.

SUMMARY

Interest in the use of exotic germplasm for improving of
cereals increased markedly following the Southern Corn Leaf
Blight epidemic of 1969-70.

Exotic germplasm varies greatly in its capacity to contribute positively to the improvement of cereals.   For this
reason, the choice of the correct exotic genotype is as important as the choice of the correct adapted breeding lines

as parents for a mating. An understanding of the evolutionary history of the species involved is essential in identi-
fying the most useful exotics.

With the possible exception of oats, the cultivated pools
of most cereals contain sufficient genetic variability to
permit further genetic gains in yield and other agronomic
traits. The gene pools of cereals could be enriched, however,
through the introgression of exotic and semi-exotic genotypes.

Exotic germplasm has long been used as a source of disease
and insect resistance. Such germplasm, if properly used,
also has the capability for increasing yields as has been
demonstrated in a number of cereals.

REFERENCES

Brown, W. L. (1975). *Proc. Ann. Corn Sorghum Res. Conf. 30*,
    81.
Browning, J. A., Frey, K. J., and Grindeland, R. L.   (1964).
    Iowa Farm Sci. 18, 5.
Carnahan, H. L. (1981). Personal communication.
Carnahan, H. L., Johnson, C. W., Tseng, S. T., and Masten-
    broek, J. J. (1978). *Crop Sci. 18*, 357.
Chang, T. T. (1979). *In* "Plant Breeding Perspectives" (J.
    Sneep and A. J. Th. Hendriksen, eds.), p. 173. PUDOC,
    Wageningen.
Chang, T. T. (1981). Personal communication.
Chang, T. T., Ou, S. H., Bathak, M. C., Ling, K. C., and
    Kaufmann, H. E. (1975). *In* "Crop Genetic Resources for
    Today and Tomorrow" (O. E. Frankel and J. G. Hawkes, eds.),
    p. 183. Cambridge Univ. Press, Cambridge.
Duvick, D. N. (1977). *Maydica 22*, 187.
Eberhart, S. A. (1971). *Crop Sci. 11*, 911.
Feldman, M., and Sears, E. R. (1981). *Sci. Amer. 244*, 102.
Frey, K. J. (1981). Personal communication.
Goodman, M. M. (1965). *Crop Sci. 5*, 87.
Griffin, B., and Lindstrom, E. W. (1954). *Agron. J. 46*,
    545.
Hallauer, A. R., and Miranda, J. B. (1981). "Quantitative
    Genetics and Maize Breeding." Iowa State Univ. Press.
Holden, J. H. W. (1976). *In* "Evolution of Crop Plants" (N.
    W. Simmonds, ed.), p. 86. Longman, London and New York.
Iltis, H. H. (1979). *Science 203*, 186.
Kwangtung Agric. and Forestry College. (1975). *Acta Genet.
    Sin. 2*, 31.
Langer, I., Frey, K. J., and Bailey, T. B. (1978). *Crop
    Sci. 18*, 938.

Lonnquist, J. H.   (1974).   *Proc. Ann. Corn Sorghum Res. Conf.*
    29, 102.
McFadden, E. S.   (1938).   *Amer. Soc. Agron. 22*, 1020.
National Academy of Sciences.   (1972).   "Genetic Vulnerabili-
    ty of Major Crops," 307 pp.   National Academy of Sciences,
    Washington, D. C.
Nelson, H. G.   (1972).   *Proc. Ann. Corn Sorghum Conf. 27*, 115.
Quinby, R., Jr.   (1974).   "Sorghum Improvement and the
    Genetics of Growth," 108 pp.   Texas A & M Univ. Press,
    College Station, TX.
Quinby, R., Jr., and Karper, R. E.   (1945).   *J. Amer. Soc.*
    Agron. 37, 916.
Reeves, R. G.   (1950).   *Agron. J. 42*, 248.
Riley, Ralph.   (1965).   *In* "Essays on Crop Plant Evolution"
    (J. B. Hutchinson, ed.), p. 103.   Cambridge Univ. Press,
    Cambridge.
Russell, W. A.   (1974).   *Proc. Ann. Corn Sorghum Res. Conf.*
    29, 81.
Seifert, R.   (1981).   Personal communication.
Tatum, L. A.   (1971).   *Science 171*, 1113.
Ting, Y.   (1933).   *Sun Yatsen Univ. Agron. Bul. 3*, 1.
Vogel, O. A., Craddock, J. C., Jr., Muir, C. E., Everson, E.
    H., and Rohde, C. R.   (1956).   *Agron. J. 48*, 76.
Vogel, O. A., Allen, R. E., and Peterson, C. J.   (1963).
    Agron. J. 55, 397.

CHAPTER 3

MULTILINE BREEDING[1]

Kenneth J. Frey

Department of Agronomy
Iowa State University
Ames, Iowa

"The multiline theory for the production of composite
varieties is one of the truly new concepts of the century in
breeding self-pollinated crops." This quotation appeared in
the Annual Report of the Rockefeller Foundation for 1962-63,
and it reflected the optimism of Borlaug (1959), who initiated
a multiline breeding program for wheat (*Triticum aestivum*) in
Mexico as a strategy for providing consistent and persistent
protection of this crop to the rust diseases (caused by *Puccinia graminis tritici* Pers. f. sp. Eriks. and Henn., *P.
glumarum* (Schm.) Eriks. and Henn., and *P. rubigo-vera tritici*
(Eriks.) Carlton). The concept of using multiline varieties
was conceived as a method for reducing the risk of losses in
crop productivity from attacks by unknown pathogens or unknown biotypes of current pathogens (Jensen, 1952).
   Natural ecosystems seldom, if ever, consist of uniform
stands of a single species, and almost certainly, never of a
single genotype of a species. Browning et al. (1979) reported
on just such a stable ecosystem in the Fertile Crescent of the
Near East. The primary annual species growing in that area
are wild barley (*Hordeum spontaneum*), wild oats (*Avena sterilis*), and wild emmer (*Triticum dicoccoides*). They studied
the *A. sterilis* interaction with several pathogens where there
is annual presence of crown rust (caused by *Puccinia coronata*

[1]*Journal Paper No. J-10309 of the Iowa Agriculture and
Home Economics Experiment Station, Ames, Iowa*

Cda. var. *avenae* Frazier and Ledingham), stem rust (caused by
P. graminis Pers. f. sp. *avenae* Eriks. and Henn.), loose smut
(caused by *Ustilago avenae* Pers. Rostr.), powdery mildew
(caused by *Erysiphe graminis avenae* El. Marchal.), and barley
yellow dwarf virus on wild oat plants. The diseases occur on
some plants, but not on others, showing that the host population is heterogeneous for resistance alleles for each pathogen. This factor of heterogeneity for pest resistance seems
to be a universal strategy for all natural plant ecosystems
resulting in a balance that predicates the survival of both
host and parasite.

Until this century, farmers used "land" cultivars, which
even though more-or-less uniform for agronomic traits, often
were heterogeneous for disease-reaction alleles. Coffman
(1977) listed 14 distinct pure-line varieties of oats (*Avena
sativa*) selected from the land cultivar 'Kherson'. He noted,
"Derivatives of Kherson differed, not only morphologically,
but also in reaction to major oat diseases, especially stem
rust."

Upon the demonstration of the simple Mendelian inheritance
of host resistance in plants (Biffen, 1905, 1912; Butler,
1905; Pole-Evans, 1911; Ward, 1905), plant selectors began to
select for homogeneous resistance in varieties, and thus was
initiated a human-guided "management system for host resistance genes." With it, both inter and intraspecific heterogeneity were lost as effective mechanisms for controlling
diseases in small grains.

A new dimension was added to breeding plants for resistance
to pests, however, when Stakman and his colleagues (Stakman
and Piemeisel, 1917; Stakeman and Levine, 1922) demonstrated
that the stem rust pathogen of wheat was composed of biotypes
(races) that were distinguishable by host-pathogen interactions. The use of homogeneous resistance, most often condi-
tioned by a single allele, in varieties of wheat, barley
(*Hordeum vulgare*), or oats grown across the whole midwestern
USA caused intense selection pressure, which led to rapid
evolution of the rust pathogen population from a race aviru-
lent on the resistance gene to a virulent one.

Small-grain varieties developed via the pure-line breeding
method quickly succumbed to the new races of the rust pathogen. As an example, even after 50 years of oat breeding in
midwestern USA for resistance to crown rust, the disease was
causing 15 to 30% loss in grain yields annually.

I. EARLY DEVELOPMENT OF MULTILINE VARIETIES

The use of planned heterogeneity in modern crop varieties
to provide stability of protection to diseases via host resistance was proposed by Rosen (1949) and refined by Jensen
(1952). Rosen (1949) proposed that an oat variety would be
the entire population of segregates from a cross, and it
would be heterogeneous for resistances to crown rust and Victoria blight (caused by *Helminthosporium victoriae* Meehan and
Murphy). Twenty years later, Rosen's concept was implemented
by Suneson (1968) when he released 'Harland' barley variety
in California, USA. This variety was the 28th generation
seed lot of a composite-cross bulk. To provide a combination
of uniformity for agronomic traits and heterogeneity for disease resistance in a common variety, Jensen (1952) proposed
the use of "multiline" varieties.

Borlaug (1959) developed a comprehensive program for producing multiline varieties of wheat. He chose four Mexican
wheat varieties as recurrent parents and backcrossed a large
number of alleles for stem-rust resistance into them to create
isolines. He proposed to blend eight to 16 isolines (each
with a different stem-rust resistance allele) into a phenotypically uniform multiline variety of wheat. No multiline
variety was released in Mexico, but 'Miramar 63' and 'Miramar
65', varieties that were heterogeneous for resistance to stem
and stripe rusts, were released in Colombia (Rockefeller
Foundation, 1963, 1965).

The multiline variety concept was first put into practice
in agricultural production on a large scale with oats in Iowa,
USA (Frey et al., 1973, 1975, 1979) more than a decade ago.
We developed two series of isolines, one with an early matur-
ing (CI 8044) and the second with a midseason (CI 7555) recurrent parent variety. The isolines in each background were
developed via backcrossing for five generations (Figure 3.1).
In our initial isoline development program, most donor parents
were unadapted to midwestern USA so crossing to Bc_5 was
critical to obtaining isolines.

Three pieces of information are used to develop an isoline
formulation for a multiline oat variety:

1. *Productivity of the isolines*. Crown-rust epidemics do
not develop every year in the midwestern USA, so a multiline
variety must produce high grain and straw yields in rust-free
as well as in rust-prevalent environments. Thus, all isolines
are tested for yield in rust-free environments. Three isolines have been inferior in yield (Frey and Browning, 1971;
Frey, 1972).

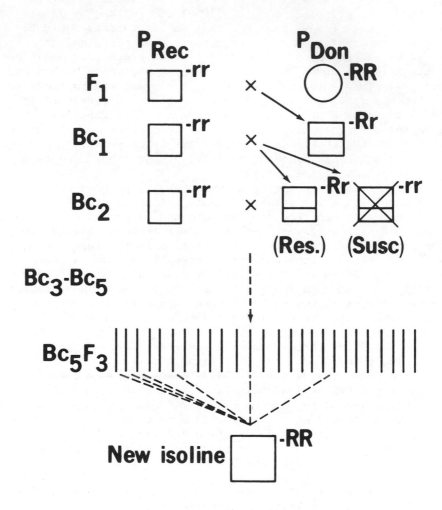

Figure 3.1. Backcrossing procedure used to develop an iso-line (Browning et al., 1964). P_{Rec} = recurrent parent; P_{Don} = donor parent; R = resistant; r = susceptible.

2. *Adult-plant reactions of isolines to races of the crown-rust pathogen.* Isolines evaluated for adult-plant reaction in single-race crown-rust nurseries in the field are rated as resistant (R), moderately resistant (MR), moderately susceptible (MS), or susceptible (S).

3. *Trends in race composition of crown-rust population.* Collections of crown-rust pustules are made each year in the USA and Canada, and these are analyzed to establish the race structure of the natural crown-rust population. These data permit the prediction of changes in proportions of races.

Information on isoline yields and adult-plant reactions to crown-rust races and the trends in race composition of the crown-rust population are used to select which isolines to use and the proportions of them to be composited to provide breeder seed of a multiline variety. The isoline composition of a typical multiline variety, 'Multiline E77', is shown in Table 3.1. It contains 10 isolines, and their percentages vary from 3 to 17. Only 63% of the plants in Multiline E77 have resistance of the R or MR type to race 326. This topic of partial susceptibility in a multiline variety will be considered later.

Each oat isoline used in a multiline variety is produced and maintained separately. Breeder seed of Multiline E77 was composited in the amount of 4.5 metric tons for the production of foundation seed. Registered and certified seed classes are produced and marketed by private growers and seed companies.

II. PRACTICAL CONCERNS WITH MULTILINE VARIETIES

As one works with multiline varieties of an autogamous crop, certain practical problems and concerns surface.

One is, "How resistant should a multiline be?" The degrees of resistance that a multiline variety should possess to the various races of a pathogen are still controversial. Borlaug (1959) proposed what Marshall (1977) called the "clean crop approach" relative to degree of resistance a multiline variety should possess. That is, all isolines in the multiline would be resistant to all prevalent races, and when an isoline showed susceptibility to a new race, it would be replaced by an alternative resistant isoline. This procedure, insofar as possible, would aim to keep the multiline disease free. There are, however, two pragmatic considerations that seem to make the clean crop approach impractical. First, to discard a gene for resistance immediately upon finding a race that is virulent upon it, causes genes to be "used up" at a fast rate. Parenthetically, the average effective life of a resistance

Table 3.1. Multiline E series of oat isolines, their pedigrees, reactions to key races of P. coronata avenae, and percentages of each in breeder seed of Multiline E77 variety.

CI number	Pedigree	Reaction to crown-rust race			Percentage in multiline E77
		294	326	264B	
9169	CI 8044^4 2 x CI 7555 x Ceirch du Bach	MR	S	MS	13
9170	CI 8044^5 3 x Clinton x Garry 2 x CI 8079^a	R	R	R	17
9172	CI 8044^6 x CI 8001^a	S	MS	MS	3
9176	CI 8044^6 x Acencao	MR	MR	S	3
9178	CI 8044^4 3 x Bonkee 2 x CI 7154 x CI 7171	R	MS	MR	13
9181	CI 8044^5 2 x Clinton x CI 8081	MR	R	R	17
9285	CI 8044^6 x PI 295919^a	MS	R	R	10
9286	CI 8044^6 x $B437^a$	R	R	R	10
9287	CI 8044^6 x $B439^a$	R	R	MR	10
9290	CI 8044^6 x $B443^a$	R	S	R	4
Percentage (R + MR)		87	63	94	--

[a] A. sterilis

gene when used alone has been 5 to 7 years for wheat and oat
rusts (Watson, 1958; Stakman and Christensen, 1959; Stevens
and Scott, 1950). Second, vertical resistance genes, by
definition, show differential reactions to races of a pathogen, and thus, if tested widely enough, no vertical gene
should show resistance to all races. So theoretically, a
"clean crop approach" probably is impossible, and certainly
impractical, to maintain.

Frey et al. (1973, 1975) and Browning and Frey (1969) have
advocated what Marshall (1977) and Marshall and Pryor (1978)
called the "dirty crop approach" with multilines of oats.
That is, no isoline in the multiline variety is completely
resistant to all races of the pathogen. "Dirty" multilines
can and will be rusted, but they are uniquely effective in
protecting the host from a pathogen by (a) stabilizing the
race structure of the pathogen population and (b) having the
resistant isolines serve as "spore traps." (Trenbath (1975)
refers to this as the "fly-paper effect.") Stabilization of
the race structure of the pathogen population should lengthen
the useful life of a resistance gene, even though more complex races may evolve on a multiline. With a stabilized
pathogen population, the spore-trapping effect reduces the
rate at which a disease builds up, resulting in a host population effect similar to horizontal resistance.

Supposedly, the validity of the "dirty-crop approach" is
dependent upon stabilizing selection (van der Plank, 1963) to
cause simple races to dominate the pathogen population.
Whether stabilizing selection actually occurs is questioned
(Flor, 1971; MacKey, 1976), but even without this phenomenon,
the "dirty" multiline should have its resistance eroded at a
slow rate.

If the "dirty-crop approach" is to be used, there still is
controversy on how much resistance a multiline variety should
possess. Browning (1974) reviewed the crown-rust resistance
data collected by Wahl (1970) and Brodny (1973) from the
indigenous ecosystem in Israel, which has a preponderance of
A. sterilis, and he concluded, "When used as part of a *diverse* population, a (resistance) gene frequency of only about
30% can be considered adequate protection against the most
virulent and prevalent group of strains of the pathogen."
Indirect evidence from mass-selection studies done by Tiyawalee and Frey (1970) indicated that 40% resistance was adequate for practical ends in Iowa, USA. Suneson (1968) suggested up to 75% resistance as desirable, and Jensen and Kent
(1963) settled on 60% resistance in New York, USA.

From both theoretical and practical standpoints, there is
good reason to use the "dirty-crop approach" for multiline
varieties. The proportion of resistance at which no signifi-

cant yield loss occurs, however, likely will vary with environmental conditions of crop production and the characteristics of the disease-causing pathogen. Intuitively, the
actual proportion of resistance in a multiline would seem to
be of less importance than its distribution among isolines.
That is, the resistance should be distributed so that each
isoline possesses both resistance and susceptibility to segments of the rust population, but no two isolines should have
similar patterns of resistance relative to pathogen races.

A second question is, "What number of components should a
multiline possess?" Browning (1974) and Parlevliet and
Zadoks (1977) argue that the greater the heterogeneity in a
multiline variety, the greater will be its durability over
time, which suggests using a very large number of components.
Practically, however, the number of isolines used in a multiline variety will be limited by the number of resistance
genes available and the work required to transfer a gene to
an isoline. Actually, to reduce the level of disease to a
point where insignificant yield losses occur requires only a
small number of isoline components (Parlevliet, 1979).

A third question posed is, "Is multiline composition
stable over generations of propagation?" This is a critical
question because (a) the effectiveness of a multiline in controlling disease might be compromised with change in its isoline composition, (b) it dictates the frequency with which
breeder seed of a multiline needs to be composited anew, and
(c) at least in the USA, to be called a "variety," a seed lot
must be stable in composition from one generation to the
next.

There is a plethora of evidence from studies on barley
(Harlan and Martini, 1938; Suneson and Wiebe, 1942; Suneson,
1949), wheat (Frankel, 1939), rice (*Oryza sativa*) (Jennings
and de Jesus, 1968), and soybeans (*Glycine max*) (Mumaw and
Weber, 1957) to show that the composition of variety mixtures
is not stable over successive generations of propagation.
Multilines, however, should not be as subject to change in
composition when grown under disease-free conditions because
they are composed of near-isolines that are nearly genotypi-
cally identical except for genes at the rust-reaction loci.
When propagated in an environment with disease presence, unstable composition might occur (Klages, 1936; Tiyawalee and
Frey, 1970), however.

Murphy et al. (1981) propagated a five-isoline version of
the Multiline M series of varieties of oats for four generations in crown-rust-free and crown-rust environments and assayed the percentages of the isolines in the multiline samples after each generation. Contrary to expectation, the
multiline, when grown under rust-free conditions, was not
stable in composition (Table 3.2). CI 9192 showed a steady

Table 3.2. Regression coefficients, deviations from regression, and mean percentages of oat isolines throughout four generations of propagation in rust-free and rusted environments (from Murphy et al., 1981).

Line of descent	Isoline	Percentage of isoline in designated generation					Regression coefficient	Variances due to deviations from regression
		0	1	2	3	4		
Rust-free	CI 9192	22	18	16	14	10	-2.80**	0.53
	CI 9183	18	23	19	19	14	-1.20	8.93
	CI 9184	20	22	31	26	38	4.00	17.07
	CI 9190	21	23	21	24	20	-0.10	3.57
	CI 9191	19	14	13	17	18	0.10	8.90
Rusted	CI 9192	22	19	19	14	15	-1.90*	2.23
	CI 9183	18	20	19	17	22	0.50	4.10
	CI 9184	20	22	25	28	28	2.20**	0.93
	CI 9190	21	23	25	24	19	-0.30	7.43
	CI 9191	19	16	12	17	16	-0.50	7.83

*Significantly different from 0.00 at the 1% level.
**Significantly different from 0.00 at the 5% level.

and significant decrease, and CI 9184 showed a consistent, although nonsignificant, increase. The other three isolines
showed no consistent changes over generations. The composi-
tional trends in the rust line of descent were less drastic,
but nevertheless similar, to those in the rust-free one. Obviously, the isolines CI 9192 and CI 9184 had weak and strong
competitive abilities, respectively, which were not associated
with rust resistance or susceptibility. Whether the changes
in multiline composition caused any degradation or enhancement
of the multiline's value for crown-rust control was not evaluated, but certainly, multilines may not be sufficiently stable
to be called varieties under USA law.

III. VALUE OF MULTILINES IN AGRICULTURAL PRODUCTION

    Multiline varieties have threefold value in agricultural
production of autogamous crops:
    1. *The pattern of virulence genes in the pathogen popula-
tion should be stabilized, relatively speaking, by the use of
multiline varieties*. Many races of the pathogen have an opportunity to survive. And because each isoline in a multiline
variety carries only one host-resistant gene, stabilizing
selection (van der Plank, 1968), if it is operational, should
aid "simple" races to predominate in the rust population.
Flor (1971), Clifford (1975), and MacKey (1976) have shown
that races of rust carry seemingly unnecessary virulence
genes. Boskovic (1976) found that most races of *P. recondita*
carried from several to 10 virulence genes even though wheat
varieties grown in the region did not possess a corresponding
number of resistance genes. Wolfe et al. (1976) concluded
that, for powdery mildew of barley, many virulence genes did
show stabilizing effects but that some did not. A similar
conclusion was given by MacKenzie (1979) for stem-rust resistance genes in wheat. On the other hand, Leonard (1969) found
evidence for stabilizing selection in the oat stem-rust pathogen, and he calculated that 35-40% of susceptibility would be
needed for each race for the simple races to become stabilized
in the rust population.
    Whether stabilizing selection is or is not a potent selec-
tion force when multiline varieties are used probably is not
of much *practical* importance in agricultural production.
Races that carry two to several virulence genes likely will
evolve, but their ability to parasitize a greater proportion
of the plants in a multiline variety will give them only marginal selective advantage. Yet, the time that would be required for a simple race to accumulate the necessary genes

for virulence and aggressiveness to become a super race, that could dominate the rust population would likely be longer than the multiline variety would be agronomically useful in agricultural production.

No information is available to predict the useful life of multiline varieties relative to pure-line varieties. Multiline oat varieties released in Iowa (Frey and Browning, 1976a, b; Frey, Browning, and Grindeland, 1971a, b) have been in production for 13 years, and there has been no detectable increase in their susceptibility to natural crown-rust populations. Admittedly though, further experience is needed before the effect of multiline varieties on the structure of the rust population, or the durability of the multiline protection, can be predicted.

2. *Multiline varieties have an effect similar to disease tolerance in delaying intrafield buildup of the pathogen.* Browning and Frey (1969), using stem rust of oats, counted uredia at rust climax on susceptible plants in large plots (30 x 30 m) sown with pure-line and multiline varieties. The study was conducted with three isolines of oats grown individually and in blends, and two rust races, also used individually or in mixtures (Table 3.3). In a 50:50 mixture of Clintland A and Clintland D inoculated with races 7 and 8 (line 7 in Table 3.3), the pustule count was only 56% as great as in pure stands of the components.

Table 3.3. Relative number of oat stem rust uredia at rust climax on susceptible culms (each value represents a mean from 80 susceptible culms) in pure stands and mixtures of Clintland isolines (Browning and Frey, 1969).

Line or mixture	Virulent race	Uredia[a]
Clintland A	8	103
Clintland D	7	97
Mean	--	100[b]
Clintland BD	None	0
Clintland A + B	8	59
Clintland D + B	7	63
Clintland A + D	7 + 8	56
Clintland A + D + BD	7 + 8	36

[a]Given as percentage of the mean of Clintland A and Clintland D grown in pure-line susceptible stands.

[b]100 percent equals 380 susceptible-type uredia per 10 culms.

Cournoyer (1967, 1970) and Browning (1974) measured yields of crown rust spores from pure-line and multiline varieties grown in large plots in Iowa (a 5-week rust season) and Texas (a 4-month rust season), respectively, USA. Plots of the first study were inoculated by transplanting seedlings infected with six crown-rust races into the center of each plot; in the second, seedlings infected with four races were transplanted at four loci in each plot; and the Texas experiment was inoculated with the natural population of the pathogen. In the first Cournoyer (1967) study, the cumulative spore load in the multiline was only about 60% as great as in the pure-line susceptible (Figure 3.2). Note that the epidemic developed to completion in the pure line, as shown by the plateau in the cumulative spore production near the end of the season. However, in the multiline, the oat plants matured normally.

In the second Cournoyer (1970) study, again the crown-rust epidemic killed the susceptible isoline prematurely (Figure 3.3). The resistant isoline and the multiline variety yielded only 6% and 25%, respectively, as many spores as the susceptible isoline.

In the Texas experiment (Browning, 1974), there were two distinct periods of rust increase on the susceptible isoline (Figure 3.4), with the midseason plateau being caused by a prolonged drought which stopped plant growth. The general trends, however, in cumulative spore production from the resistant and susceptible isolines and the multiline variety were similar in Texas and Iowa.

All three of these experiments show that the multiline was effective in holding the crown rust in check. It is postulated that this multiline effect occurs because of "spore trapping" or the "fly-paper effect." That is, in the initial spore shower that inoculates a field of multiline oats, a spore that lands on a resistant plant cannot reproduce: This serves to decrease the amount of initial effective inoculum. Likewise, progeny spores from the initial infections on susceptible plants in the multiline cannot infect and reproduce if they land on resistant plants: This reduces the rate at which the already established pathogen can increase.

According to van der Plank (1963), the rapidity with which an epidemic of a crowd disease, at least in the early stages, develops in a host population is described by the equation

$$X_t = X_o e^{rt}$$

where X_t = accumulated spore (pustule) number at time t, X_o = initial effective inoculum, r = rate of increase in spore (pustule) per day, and e = 2.718. Utilizing crown rust data on oats, Frey et al. (1973) and Browning and Frey (1969)

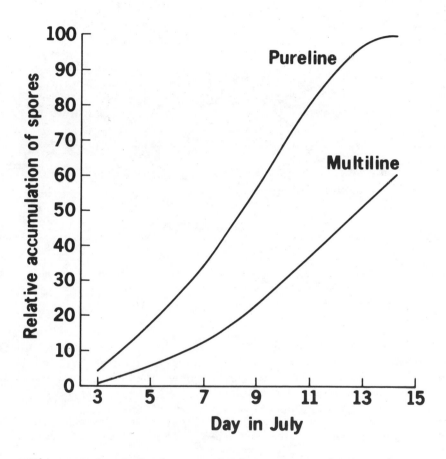

Figure 3.2. Relative cumulative counts of crown-rust spores collected outside of pure-line and multiline plots of oats inoculated with six races, Ames, Iowa (adapted from Cournoyer, 1967).

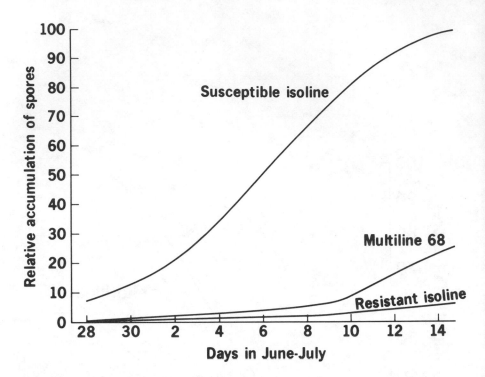

Figure 3.3. Relative cumulative counts of crown-rust spores collected outside of plots sown to susceptible and resistant isolines and a commercial multiline variety of oats that had been inoculated with four races, Ames, Iowa (adapted from Cournoyer, 1970).

hypothesized the effects of various resistance and variety types on X_o and r (Table 3.4). A pure-line variety with one vertical resistance gene reduces X_o, but has no effect on r. A tolerant variety, oppositely, has no effect on X_o, but the spore yield per pustule and, therefore, r is reduced. The multiline, however, reduced both X_o and r because of spore trapping.

Gill et al. (1979) reported on the ranges of rust development on plants in a multiline and in the isoline components when grown in pure stands (Table 3.5). To quote the authors, "It is clearly evident . . . that in multilines the rust development is much lower as compared with the development on pure cultures of the component lines."

Luthra and Rao (1979) estimated X_o's and r's for leaf-rust infections on seven multiline varieties of wheat and the means for the components when grown in pure stands. The

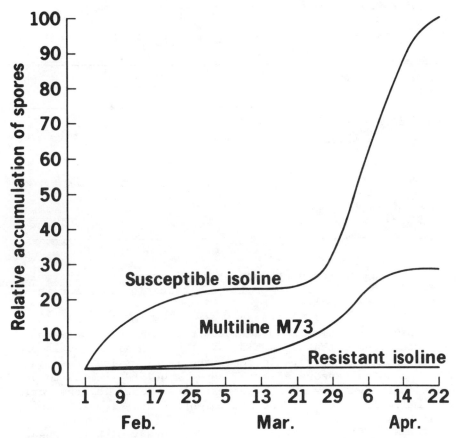

Figure 3.4. Relative cumulative counts of crown-rust spores collected outside of plots sown to susceptible and resistant isolines and a commercial multiline variety of oats that had natural disease infections, Rolestown, Texas (Browning, 1974).

recurrent parent was 'Kalyansona', and 12 races of leaf rust were used to inoculate the spreader rows in the experiment. X_o was estimated by counting the pustules after initial infection occurred, and r was estimated from leaf-rust ratings made at five periods ca. a week apart. The reduction in X_o, relative to the mean of the components grown in pure stand, ranged from 45% for MLKS-7506 to 76% for MLKS-7507 (Table 3.6). There was no relationship between reduction in X_o and either the number of components or proportion of resistance in the multiline. In five of six infected multilines, r, relative to the mean of the components, was reduced from 3 to

Table 3.4. *Effects of vertical resistance and tolerance to crown rust upon level of initial inoculum (X_o) and rate of disease increase (r) in pure-line and multiline oat cultivars (Frey et al., 1973).*

Resistance type in cultivar	Effect on	
	X_o	r
Vertical resistance in pure line	Reduced	None
Tolerance in pure line	None	Reduced
Vertical resistance in multiline	Reduced	Reduced

16%, whereas in the sixth one, r was 4% greater in the multi-line (Table 3.7). The overall mean reduction in r for the multilines was 4%.

Luthra and Rao (1979) used their values of X_o and r in Kalyansona and multilines derived from them to compute the numbers of days that it would take to reach given levels of infection (Table 3.8). To reach the level of leaf-rust infection on Kalyansona after 31 days, required from 39 to 50 days in the various multiline varieties. That is, the various multilines delayed the attainment of this level of infection from 8 to 19 days.

Fried et al. (1979a, b) conducted a study with the wheat-powdery mildew (*Eriysiphe graminis* f. sp. *tritici*) system to determine the effect of different ratios of resistance to susceptibility upon the effective initial inoculum (X_o) and the apparent rate of disease increase (r). They used 'Chancellor' wheat variety (susceptible to mildew) alone and with a resistant near isoline in proportions of 1:1, 1:2, and 1:3 susceptible to resistance. The variety and three blends were sown in a replicated experiment in widely separated plots 9 x 9 m. Infection was either from naturally occurring inoculum or from actively sporulating transplants. The estimates of disease incidence were made via visual ratings. X_o was reduced significantly by levels of resistance from 50 to 75% at both the 15- and 23-day reading periods (Table 3.9). Rate of apparent increase (r) was not reduced significantly, however, until the plant population had 75% resistance. And, even at the 75% level, the reduction in r was only 8%.

All studies conducted on the subject show that airborne diseases develop more slowly on multiline varieties of wheat and

Table 3.5. Ranges of reactions of yellow (P. striiformis) and brown rust (P. recondita) on plants of seven Kalyansona and seven P.V. multiline varieties of wheat and their component isolines grown in pure stands in India (Gill et al., 1979).

Multiline	Range of rust reaction to			
	Yellow rust in		Brown rust in	
	Multiline	Components	Multiline	Components
Kalyansona multilines				
KSML1	0-ts	0-50S	0	0-20S
KSML2	0-ts	0-15S	0	0-30S
KSML3	0	0-5S	0	0
KSML4	0-ts	0-10S	0	0
KSML5	0-ts	0-ts	0	0
KSML6	0-ts	0-5S	0	0-5S
KSML7	0	0-10S	0-ts	0-5S
PV multilines				
PVML1	0	0-20S	t	0-5MR
PVML2	0	0-20S	0	0-20S
PVML3	0	0-20S	0	0-trS
PVML4	0-ts	0-10S	0	0-15S
PVML5	0	0-20S	0-ts	0-15S
PVML6	0-ts	0-20S	0-ts	0-30S
PVML7	0	0-30S	0-ts	0-10S

Table 3.6. Number of components and ratio of resistance (R) to susceptibility (S) in and percentage reduction in X of leaf rust, relative to its components, for seven Kalyansona multiline varieties of wheat (adapted from Luthra and Rao, 1979).

Multiline	Number of components	Ratio R:S	Percent reduction in X_o
MLKS-7501	4	50:50	74
MLKS-7502	5	40:60	68
MLKS-7503	5	30:70	72
MLKS-7504	3	100:0	--
MLKS-7505	7	60:40	50
MLKS-7506	7	72:28	45
MLKS-7507	12	42:58	76

Table 3.7. Rates (r) of increase for leaf rust infection for multiline varieties of wheat and mean r's for their components and percent reductions in r, relative to means of components, due to multiline effect (adapted from Luthra and Rao, 1979).

Multiline	Rate (r) of infection for		Percent reduction in r in multiline
	Multiline	Component mean	
MLKS-7501	0.2692	0.2970	9
MLKS-7502	0.3005	0.3091	3
MLKS-7503	0.3072	0.3177	3
MLKS-7504	No infection		
MLKS-7505	0.2512	0.2970	15
MLKS-7506	0.2493	0.2970	16
MLKS-7507	0.2788	0.2679	-4
Mean	0.2860	0.2976	4

oats than they do on pure lines. This slower development is due primarily to multilines reducing the effective initial inoculum (X_o), but in some instances, the rate of infection (r) also causes reduced buildup. The practical result is that plants in a "dirty" multiline (each plant of which is susceptible to some component in the rust population), because they "rust" more slowly than when the component isolines are grown in pure stands, produce more biomass and grain per culm (on a per unit area) than the means of their counterparts in pure

Table 3.8. Numbers of days to reach a given level of leaf rust infection in Kalyansona wheat variety and seven multilines derived from it, and days of delay in attaining the infection level due to reductions in X_o and r in multilines (adapted from Luthra and Rao, 1979).

Variety or multiline	Days to given rust level	Delay due to reduction in X_o	r	Total delay
Kalyansona	31^a	--	--	--
MLKS-7501	47	8.5	7.5	16
MLKS-7502	40	6.0	3.0	9
MLKS-7503	39	5.5	2.5	8
MLKS-7504	No infection			
MLKS-7505	48	7.0	10.0	17
MLKS-7506	50	8.0	11.0	19
MLKS-7507	46	9.0	6.0	15

[a] *A time of 31 days (5 March–5 April) and r = 0.3333 per unit per day were used for Kalyansona.*

Table 3.9. Effective initial inoculum (X_a) and rate of apparent increase (r) values for powdery mildew in populations of wheat with different proportions of susceptible to resistant plants (adapted from Fried, et al., 1979a, b).

Host	X_a values y day after inoculation $y = 15$	$y = 23$	r
Cc^a	$3.7\ a^b$	3.9 a	0.22 a
1:1	3.2 ab	3.3 b	0.22 a
1:2	2.9 bc	3.3 b	0.21 ab
1:3	2.5 c	3.0 b	0.20 b

[a] *Cc = Chancellor*

[b] *Values in column with same letter are in same significance interval.*

stands. A secondary factor that increases yield may be the ability of less-infected plants to compensate for heavily infected ones by their utilization of additional environmental factors (e.g., sunlight, moisture) that are not utilized by the heavily infected ones.

   3. *Multiline varieties provide a strategy to insure that
genetic heterogeneity is present in the arena of agricultural
production.* A National Academy of Sciences (NAS) report
(1972) concludes that (a) vulnerability of major agricultural
USA crops to pests and major climatic fluctuations stems from
genetic uniformity and (b) some American crops are, on this
basis, highly vulnerable. A good illustration of point (a)
was the loss in corn (*Zea mays*) production that occurred in
the USA in 1970 due to an epidemic of southern corn leaf
blight disease (caused by *Helminthosporium maydis* race T)
(Tatum, 1971). Practically the entire corn acreage in the
USA was sown with hybrids that possessed a single cytoplasm,
T-type male sterile, which caused susceptibility to this
disease, and resulted in a national loss of 15%. To illustrate (b), the NAS report showed that, in 1969, three major
varieties accounted for 53% of the USA cotton (*Gossypium hirsutum*) acreage, six varieties accounted for 71% of the corn
acreage, and six varieties occupied 56% of the soybean acreage. Truly, in the developed countries of the world, most
major crops are produced with a very small number of varieties, and this phenomenon is rapidly coming about in many
developing countries also.
   Two factors lead to only a few varieties being used for
any crop. First, the market for some crops demands uniformity
(e.g., milling quality in wheat), and only a few varieties
will meet this market standard at any point in time. Second,
varieties have different productivities and characteristics,
and very soon, even though many varieties may be available,
the agricultural production community (farmers) will have reduced the effective number in use to the "best" few.
   Multiline varieties, if adopted as a general plant breeding strategy for a crop, would be one way to keep genetic
heterogeneity in the agricultural production arena regardless
of whatever selection among varieties may be done by the pro-
ducers or caused by market demand.

IV. COMBINING MULTILINES WITH OTHER DISEASE-CONTROL
     STRATEGIES

   The case has been made that the multiline strategy for
managing host resistance genes can be effective for increasing the durability of a variety and to reduce disease buildup
in production fields. However, oat researchers in the "Puccinia Path" of North America have combined it with several
other strategies into a total management system to control
losses from crown rust (Frey et al., 1973, 1979).
   Breeders and pathologists recognize that the crown rust

organism has evolved into a "continental" pathogen, which can rove up and down the Puccinia Path of North America to attack oats whenever its growing season occurs (Figure 3.5) (Browning et al., 1969). The Puccinia Path has three regions relative to movement of the crown-rust pathogen: (a) the southern region, where the pathogen overwinters; (b) the central region, which serves as a transmittal zone from south to north and vice versa; and (c) the northern region, where oats is a full-season crop and spore showers occur when oat plants are in the juvenile stage. A plan has been worked out to force the crown-rust pathogen to recycle between incompatible oat varieties (i.e., varieties with different genes for resistance) in the overwintering and oversummering areas by deploying different sets of resistance genes for the three regions. Each set was composed so that no known race of crown rust would be virulent on a resistance gene in more than one region.

The oat workers in Iowa also use a type of disease resistance known as "tolerance" to reduce losses due to crown rust on oats. Caldwell et al. (1958) described tolerance of small grains to rust disease as "disease endurance," and Browning and Frey (1969) stated that, "Tolerant plants look susceptible but yield resistant." Because the infection symptoms on tolerant plants are rated susceptible, tolerance must be measured via yield reduction caused by an attack of rust. Tolerance indexes are computed with the formula

$$\text{Tolerance index} = \frac{\text{Grain yield in plot with rust}}{\text{Grain yield in plot without rust}}$$

This trait is quantitatively inherited with a heritability of 67 to 84% (Simons, 1969; Simons and Michel, 1967). In Iowa, oat genotypes used as recurrent parents to develop multiline varieties must possess tolerance.

Thus, for crown rust control in the Puccinia Path in North America, three strategies or components are being used in a total management system to control crown rust. First, genes for vertical resistance are deployed on a regional basis to provide interregional diversity. Second, the set of genes assigned to the central region is being used to develop multiline varieties. Third, genotypes used as recurrent parents must have tolerance to crown rust. It also is conceivable that other strategies in crop production (e.g., occasional use of fungicides, crop rotation) could be incorporated with management strategies for host resistance to give integrated pest management.

Figure 3.5. Puccinia Path of North America and the three regions into which genes for vertical resistance to crown rust are deployed (Frey et al., 1973).

V. MULTILINE BREEDING IN 1981

    It has been nearly three decades since Jensen (1952) pre-
sented the arguments and plan for developing multiline varieties of small grains, but recently, Browning et al (1979)
stated, "Even with such well justified enthusiasm for the
multiline concept, the actual use of multiline varieties has
not caught on." Recent evidence shows, however, that in fact,
multiline breeding or some modification "has caught on" by
1981.
    The multiline breeding program for stem rust resistance
that Borlaug (1959) mounted for wheat in the 1950's never resulted in the release of multiline varieties for production
in Mexico because new semidwarf varieties (e.g., 'Pitic 62')
that would yield twice as much as the native Mexican varieties
were ready for release also (Rajaram and Dubin, 1977). In the
early 1970's, the CIMMYT (Centro Internacional para Mejoramiento de Maiz y Trigo) wheat program again began a multiline-
breeding project for resistance to stem, leaf, and stripe
rusts using the semidwarf 'Siete Cerros' variety as the recurrent parent, and many lines and varieties with genes for
resistance to stem, leaf, and stripe rust as donor parents.
They intermate the single crosses into double crosses of the
type (A x B) x (A x C), where A = Siete Cerros and B and C
represent donor parents. Segregates that are phenotypically
like Siete Cerros and carry rust resistance are candidates
for components in a multiline variety. Many of these candi-
date lines along with experimental multilines from the CIMMYT
program have been tested in international nurseries since 1975
(Rajaram and Dubin, 1977; Rajaram et al., 1979).
    In the Netherlands, a multiline variety of wheat 'Tumult'
with resistance to stripe rust has been tested extensively
since 1974 (Groenewegen, 1977; Groenewegen and Zadoks, 1979).
These authors state, "In 1977, yellow rust was found at heading time in almost all 'Tumult' multiplication fields (60 ha).
. . . In all multiplications fields of 'Tumult' the infection
remained at the trace level."
    Gill et al., (1979, 1980), in India, have been experiment-
ing with multiline varieties of the Kalyansona (Siete Cerros)
type with resistance to stripe rust since 1974. Recently,
one of these varieties, KSML-3, has been released for use by
farmers in the Punjab wheat-producing area of India (Borlaug,
1981). Multiline varieties of wheat with resistance to leaf,
stem, and stripe rust are being researched at Pantnagar, India
by Gautam et al. (1979) and, for resistance to brown rust, at
Hissar by Chowdhury et al. (1979). MLKS-11 and KML 7406 multiline varieties have been released for use in the wheat-

growing areas around New Delhi and Kanpur, India (pers. commun. from K. S. Gill, Dept. Plant Breeding, Pubjab Agric.
Univ., Ludhiana, India).

In Washington, USA, a multiline variety of wheat with resistance to stripe rust is scheduled for release in 1982
(pers. commun. from R. E. Allen, Agron. Dept., Washington
State Univ., Pullman, Wash.). It is a club-type semidwarf
wheat with 10 components that carry nine sources of resistance
to stripe rust and four to leaf rust. In 120 comparisons, it
has yielded 3% more than the mean of its components.

Wolfe (1977) and Wolfe and Barrett (1977, 1980) are using
variety mixtures of barley to control powdery mildew disease
(*Erysiphe graminis hordei* DC).

The isolines used in the multiline oat varieties developed
and released in Iowa, USA, in Tumult wheat variety in The
Netherlands, and in the multiline wheat variety to be released
in Washington, USA, were developed with several generations of
backcrossing; thus, they tend to conform to the Jensen model.
Some plant breeders have criticized this strict approach to
multiline variety development: (a) It takes 2 to 5 years of
backcrossing, selecting, and increasing seed to convert a good
genotype to a multiline (Groenewegen and Zadoks, 1979). The
argument is that, by the time that a good genotype can be converted, a better and higher-producing variety will be available for release (Wolfe and Barrett, 1980). For example, the
development of the high-yielding semidwarf wheats in the 1958-
62 era kept the first Mexican multilines from being released.
The validity of this criticism fits only certain cases, and
then only in certain areas. (b) Strictly bred multiline
varieties usually have only one target disease, and their usefulness in many situations may be limited by nontarget diseases (Groenewegen and Zadoks, 1979). Of course, a number of
multiline varieties have been developed with simultaneous
heterogeneity of resistance to several diseases (Rajaram and
Dubin, 1977; Gill et al., 1979).

Wolfe and Barrett (1977, 1980) have proposed to alleviate
these two criticisms of the multiline approach to breeding by
using variety mixtures instead of isoline mixtures. They have
worked exclusively with malting barley and powdery mildew.
The strategy used is to mix three adapted malting varieties of
barley, each of which has a different gene(s) for resistance
to some prevalent races of the mildew pathogen and which are
somewhat diverse in agronomic characteristics. The authors
claim that variety mixtures have two pathological and one agronomic advantage over traditional multiline varieties.

The first proposed pathological advantage for a variety
mixture is that it would cause disruptive selection for fitness genes. A major concern is that a multiline will provide

an optimum breeding ground for a "super" race of the pathogen;
i.e., a biotype with enough virulence genes to attack all resistance genes in the host population. Wolfe and Barrett
(1977) and MacKey (1976) point out, however, that there are
really two components to the development of a successful
super race; i.e., one which could devastate a multiline variety. One component requires that the race accumulate the
necessary virulence genes to permit it to parasitize all resistance genes in the multiline host. The pathogen genotype
with all of the virulence genes needed to be a super race will
not predominate in the pathogen population, however, unless
it also has the proper "fitness" genes to grow vigorously in
its physical and biological environment. An important element
of that environment is the genotype of the host on which the
pathogen is obligated to grow and reproduce. The host genes
that interact with the fitness genes of the pathogen would be
quite uniform for a multiline because it is composed of isolines. Thus, evolution of a pathogen fitness genotype compatible with the multiline host genotype would be persistently
directional. The variety mixture would present the same problem to the pathogen as would a multiline relative to accumu-
lating virulence genes to become a super race, but it would
be different from the multiline in that it would present a
variable biological environment to the pathogen relative to
evolution of fitness because each host variety would represent a different genotypic environment with respect to fitness
of the pathogen. Therefore, the pathogen would be subjected
to disruptive selection for fitness, especially if varieties
in the mixture are changed often.
  The second proposed pathological advantage of variety
mixtures over multilines is related to nontarget diseases
(Groenewegen and Zadoks, 1979). Most multiline varieties
bred to date have had one primary disease for which the multiline is targeted; e.g., multiline varieties of oats in Iowa
are targeted to control only crown rust. Because the re-
mainder of the multiline genotype is uniform, the multiline
variety is as vulnerable to nontargeted pathogens as is a
pure-line variety. A good example of a nontargeted disease
that wiped out a crop was Victoria blight (caused by *Helminthosporium victoriae*) that severely attacked oat varieties
that had been bred to carry the Victoria gene for crown-rust
resistance (Murphy and Meehan, 1946). Wolfe and Barrett
(1980) have no experience with nontarget diseases on their
barley mixtures, but Groenewegen and Zadoks' (1979) variety
mixtures would be much superior to multilines for controlling
nontargeted diseases. Of course, several multiline breeding
programs are designed to produce "modified" multilines that
are targeted against several diseases (Rajaram et al., 1979).

One proposed agronomic advantage of variety mixtures over multilines is that it permits breeders to move new varieties with superior agronomic traits and yielding ability into agricultural production more rapidly than if they must be converted into multilines. Groenewegen and Zadoks (1979) estimate this time saving to be from 2 to 5 years.

Wolfe and Barrett (1980) propose the second agronomic advantage of variety mixtures to be the yield synergism that results. Quite commonly, a 2-5% yield advantage results from sowing variety mixtures. These authors found that their barley variety mixtures produced 3% more yield than the mean of the varieties grown in pure stand when no mildew was present and a 9% advantage when mildew was epidemic.

The primary deterrent to growing variety mixtures is whether the consuming market will accept a potentially variable product. If used as feed for livestock, grain uniformity would be of no consequence, but it is possible that variable milling quality in wheat would not be acceptable. Wolfe and Barrett (1980) have found maltsters willing to buy and process seed lots harvested from a variety mixture, however.

Probably the ultimate use of intravarietal heterogeneity to control diseases is apt to settle with "mixtures of related lines" rather than variety mixtures or classical multilines.

SUMMARY

Multiline varieties of crop plants are useful in controlling certain diseases. They provide durability of disease resistance, reduce disease buildup in fields of crops, and provide a mechanism whereby genetic diversity can be maintained in the agricultural production setting. The mechanism by which the disease buildup in a field is slowed seems to be via "spore trapping" by resistant plants. This reduces the effective initial inoculum (X_o) and the rate of infection (r).

Multiline composition may not be stable, so they may need to be recomposited annually. Generally, it is of value to use the multiline strategy in combination with other strategies, such as disease tolerance and resistance-gene deployment, to give effective integrated disease control.

Multiline variety breeding programs now exist at CIMMYT, Wageningen, The Netherlands, at three research stations in India, and in Iowa and Washington, USA. Perhaps variety mixtures or multilines of similar but not isolines, while providing the heterogeneity for disease control, will be better than strict multilines as a defense against nontarget diseases and to permit the use of agronomic improvements in agricultural production more rapidly.

REFERENCES

Biffen, R. H.   (1912).   *J. Agric. Sci. 4*, 421.
Biffen, R. H.   (1905).   *J. Agric. Sci. 1*, 4.
Borlaug, N. E.   (1981).   *In* "Plant Breeding II" (K. J. Frey, ed.), p. 473. Iowa State Univ. Press, Ames, Iowa.
Borlaug, N. E.   (1959).   *Proc. First Int. Wheat Genet. Symp.*, p. 12. Winnipeg, Canada.
Boskovic, M. M.   (1976).   *Proc. 4th Eur. Meditt. Cereal Rust Conf.*, p. 75. Interlaken, Switzerland.
Brodny, U.   (1973). M.S. Thesis, Tel Aviv Univ., Tel Aviv, Israel.
Browning, J. A.   (1974).   *Proc. Am. Phytopathol. Soc. 1*, 191.
Browning, J. A., and Frey, K. J.   (1969).   *Annu. Rev. Phytopathol. 7*, 355.
Browning, J. A., Frey, K. J., and Grindeland, R.   (1964). *Iowa Farm Sci. 18*(8), 5.
Browning, J. A., Frey, K. J., McDaniel, M. E., Simons, M. D., and Wahl, I.   (1979).   *Indian J. Genet. Plant Breed. 39*, 3.
Browning, J. A., Simons, M. D., Frey, K. J., and Murphy, H. C. (1969).   *In* "Disease consequences of intensive and extensive culture field crops" (J. A. Browning, ed.), p. 49. *Iowa Agric. Home Econ. Exp. Stn. Rep. 64*.
Butler, E. J.   (1905).   *J. Agric. Sci. 1*, 361.
Caldwell, R. M., Schafer, J. F., Compton, L. E., and Patterson, F. L.   (1958).   *Science 128*, 714.
Clifford, B. C.   (1975).   *Rep. Welsh Pl. Breed. Stn. 1974*, 107.
Chowdhury, R. K., Singh, R. K., Yadava, R. K., and Chowdhury, J. B.   (1979).   *Indian J. Genet. Plant Breed. 39*, 95.
Coffman, F. A.   (1977).   *U.S. Dep. Agric. Tech. Bul. 1516*. Washington, D. C.
Cournoyer, B. M.   (1970). Ph.D. Dissertation, Iowa State Univ., Ames, Iowa. Abstract number 70-25778.
Cournoyer, B. M.   (1967). M.S. Thesis, Iowa State Univ., Ames, Iowa.
Flor H. H.   (1971).   *Annu. Rev. Phytopathol. 9*, 275.
Frankel, O. H.   (1939).   *J. Agric. Sci. 29*, 249.
Fried, P. M., MacKenzie, D. R., and Nelson, R. R.   (1979a). *Phytopathol. Z. 95*, 140.
Fried, P. M., MacKenzie, D. R., and Nelson, R. R.   (1979b). *Phytopathol. Z. 95*, 151.
Frey, K. J.   (1972).   *Crop Sci. 12*, 809.
Frey, K. J., and Browning, J. A.   (1976a).   *Crop Sci. 16*, 311.
Frey, K. J., and Browning, J. A.   (1976b).   *Crop Sci. 16*, 312.
Frey, K. J., and Browning, J. A.   (1971).   *Crop Sci. 11*, 757.
Frey, K. J., Browning, J. A., and Grindeland, R. L.   (1971a). *Crop Sci. 11*, 939.

Frey, K. J., Browning, J. A., and Grindeland, R. L. (1971b).
  Crop Sci. 11, 940.
Frey, K. J., Browning, J. A., and Simons, M. D. (1979).
  Indian J. Genet. Plant Breed. 39, 10.
Frey, K. J., Browning, J. A., and Simons, M. D. (1975).
  Sabrao J. 7, 113.
Frey, K. J., Browning, J. A., and Simons, M. D. (1973).
  Z. Pflanzenkr. Pflanzenschutz. 80, 160.
Gautman, P. L., Malik, S. K., Pal, S., and Singh, T. B.
  (1979). *Indian J. Genet. Plant Breed., 39*, 72.
Gill, K. S., Nanda, G. S., Singh, G., and Aujula, S. S.
  (1979). *Indian J. Genet. Plant Breed. 39*, 30.
Gill, K. S., Nanda, G. S., Singh, G., and Aujula, S. S.
  (1980). *Euphytica 29*, 125.
Groenewegen, L. J. M. (1977). *Cereal Res. Commun. 5*, 125.
Groenewegen, L. J. M., and Zadoks, J. C. (1979). *Indian J.
  Genet. Plant Breed. 39*, 81.
Harlan, H. V., and Martini, M. L. (1938). *J. Agric. Res. 57*,
  189.
Jennings, P. R., and de Jesus, J. (1968). *Evolution 22*, 119.
Jensen, N. F. (1952). *Agron. J. 44*, 30.
Jensen, N. F., and Kent, G. C. (1963). *Farm Res. 29(2)*, 4.
Klages, K. H. (1936). *J. Am. Soc. Agron. 28*, 935.
Leonard, K. J. (1969). *Phytopathology 59*, 1851.
Luthra, J. K., and Rao, M. V. (1979). *Indian J. Genet.
  Plant Breed. 39*, 38.
MacKenzie, D. R. (1979). *Proc. Rice Blast Workshop.*, p. 199.
  Rice Res. Inst., Los Banos, Philippines
MacKey, J. (1976). *Proc. 4th Eur. Meditt. Cereal Rust. Conf.*,
  p. 35. Interlaken, Switzerland.
Marshall, D. R. (1977). *Ann. N.Y. Acad. Sci. 287*, 1.
Marshall, D. R., and Pryor, J. R. (1978). *Theor. Appl.
  Genet. 51*, 177.
Mumaw, C. R., and Weber, C. R. (1957). *Agron. J. 49*, 154.
Murphy, H. C., and Meehan, F. (1946). *Phytopathology 36*,
  407.
Murphy, J. P., Helsel, D. B., Elliott, A., Thro, A. M., and
  Frey, K. J. (1981). *Euphytica 30*, (in press).
National Academy of Sciences. (1972). *NAS-NRC Publ.* ISBNO-
  309-02030-1.
Parlevliet, J. E. (1979). *Indian J. Genet. Plant Breed. 39*,
  22.
Parlevliet, J. E., and Zadoks, J. C. (1977). *Euphytica 26*,
  5.
Pole-Evans, I. B. (1911). *J. Agric. Sci. 4*, 95.
Rajaram, S., and Dubin, H. J. (1977). *Ann. N.Y. Acad. Sci.
  287*, 243.

Rajaram, S., Sokvmand, B., Dubin, H. J., Torres, E., Anderson, R. G., Roelfs, A. T., Samborski, B. J., and Watson, I. A. (1979). *Indian J. Genet. Plant Breed. 39*, 60.

Rockefeller Foundation. (1965). *Annu. Rep. 1964-65.*

Rockefeller Foundation. (1963). *Annu. Rep. 1962-63.*

Rosen, H. R. (1949). *Phytopathology 39*, 20.

Simons, M. D. (1969). *Phytopathology 59*, 1329.

Simons, M. D., and Michel, L. D. (1967). *Plant Dis. Rep. 51*, 130.

Stakman, E. C., and Christensen, J. J. (1959). *In* "Plant Pathology: An advanced treatise" (J. J. Horsfall and A. E. Diamond, eds.), *3*, 567.

Stakman, E. C., and Levine, N. N. (1922). *Minn. Agric. Exp. Stn. Tech. Bul. 8.*

Stakman, E. C., and Piemeisel, P. J. (1917). *Phytopathology 7*, 73.

Stevens, N. E., and Scott, W. O. (1950). *Agron. J. 42*, 307.

Suneson, C. A. (1968). *Calif. Agric.* (August 1968). p. 9.

Suneson, C. A. (1949). *Agron. J. 41*, 459.

Suneson, C. A., and Wiebe, G. A. (1942). *J. Am. Soc. Agron. 24*, 1052.

Tatum, L. A. (1971). *Science 171*, 1113.

Tiyawalee, D., and Frey, K. J. (1970). *Iowa State J. Sci. 45*, 217.

Trenbath, D. R. (1975). *Ecologist 5*, 76.

van der Plank, J. E. (1968). "Disease Resistance in Plants." Academic Press. New York and London.

van der Plank, J. E. (1963). "Plant Diseases: Epidemics and Control." Academic Press. New York.

Wahl, I. (1970). *Phytopathology 60*, 746.

Ward, H. M. (1905). *Ann. Bot. 19*, 1.

Watson, I. A. (1958). *Agric. Gaz., N.S.W. Dep. Agric. 69*, 630.

Wolfe, M. S. (1977). *Cereal Res. Commun. 5*, 111.

Wolfe, M. S., and Barrett, J. A. (1980). *Plant Dis. Rep. 64*, 148.

Wolfe, M. S., and Barrett, J. A. (1977). *Ann. N.Y. Acad. Sci. 287*, 151.

Wolfe, M. S., Barrett, J. A., Shattock, R. C., Shaw, B. F., and Whitbread, R. (1976). *Ann. Appl. Biol. 82*, 369,

CHAPTER 4

BREEDING CROP PLANTS
TO TOLERATE SOIL STRESSES

F. N. Ponnamperuma

The International Rice Research Institute
Los Baños, Laguna, Philippines

Plant breeding has occurred in nature's laboratory for
about 3 billion years and in human habitats for the past
6,000 years. That process has provided genetic diversity in
the form of germplasm collections that have been used to
breed crop plants for high yield, quality, and resistance to
diseases and insects. But little was done until about a
decade ago to select and breed crop plants for tolerance to
soil stresses.

Four recent developments have stimulated interest in
breeding crop plants for tolerance to soil stresses:

1. the need for more food for a rapidly expanding population;
2. the scarcity of arable land;
3. the limitations of modern crop technology; and
4. the high cost of energy.

The food deficit problem can be solved by extending the
area of cultivated land and increasing the yield per ha per
year. In either case, there are problems. Because good
arable land is scarce, the area in crop production can be increased only by moving production onto marginal lands that
lie idle because of soil stresses. Besides 5 to 7 million ha
of arable land deteriorate into marginal land every year
(Dudal, 1978). The problem lands with stresses can be reclaimed by soil and water management or by chemical amendments, but reclamation requires costly initial and recurrent

inputs. Use of tolerant cultivars may enable cultivation of soils where stresses are not severe without reclamation measures, and also lower the cost of reclamation on land where the stresses are moderately severe.

The principal obstacles to increasing yield per ha per year on current cultivated land are soil stresses and the high cost of fertilizers. Again, the use of soil amendments and fertilizers may be minimized by using cultivars that tolerate soil stresses.

I. SOIL STRESSES

Soil stresses fall into three classes: chemical (mainly mineral), physical, and hydrological. This paper will discuss mineral stresses.

According to Dudal (1976), soils with mineral stresses cover 3 billion ha or 22% of the world's land surface. Mineral stresses that depress crop yields are either soil toxicities or nutrient deficiencies or both. Toxicities occur in saline soils, sodic soils, strongly acid soils, acid sulfate soils, and peat soils. Nutrient deficiencies occur in most soil classes.

In Southern and Southeastern Asia, where both food and arable land are scarce, about 100 million ha of land that is climatically, physiographically, and hydrologically suited to rice production lie uncultivated largely because of soil stresses (Table 4.1)

Saline Soils

Extent and distribution. Saline soils cover over 380 million ha, of which 240 million ha are not strongly saline (Massoud, 1974) and have crop production possibilities. About 60 million ha in Southern and Southeastern Asia of which 27 million ha in the humid tropics are climatically, physiographically, and hydrologically suited to rice but are not cultivated because of salinity (Akbar and Ponnamperuma, 1980).

Characteristics. Saline soils contain sufficient salts in the root zone to impair growth of crop plants. Current definitions of saline soils are based on salt content alone or in conjunction with texture, morphology, or hydrology (Northcote and Skene, 1972; FAO/UNESCO, 1973; FAO, 1974; United States Salinity Laboratory Staff, 1974; Soil Science Society

Table 4.1. Distribution and extent of problem rice lands
in South and Southeast Asia.

Country	Saline soils	Alkali soils	Acid sulfate soils (millions of ha)	Peat soils	Total
Bangladesh	2.5	0.5	0.7	0.8	4.5
Burma	0.6		0.2		0.8
India	23.2	2.5	0.4		26.1
Indonesia	13.2		2.0	16.0	31.2
Kampuchea	1.3		0.2		1.5
Malaysia	4.6		0.2	2.4	7.2
Pakistan	10.5	4.0			14.5
Philippines	0.4				0.4
Thailand	1.5		0.6	0.2	2.3
Vietnam	1.0		1.0	1.5	3.5
Total	58.8	7.0	5.3	20.9	92.0

Source: International Rice Research Institute, 1980.

of America, 1978). The most widely accepted definition of a
saline soil is one that gives an electrical conductivity in
the saturation extract (EC$_e$) exceeding 4 mmho/cm at 25°C.
However, saline soils vary widely in their salt source and
chemical, physical, and hydrological properties (Ponnamperuma
and Bandyopadhya, 1980).

The salt source may be seawater, surface or underground
saline water, or wind. Dominant salts in coastal soils are
chlorides, whereas in inland areas chloride sulfates predominate. The salt content varies horizontally, vertically, and
seasonally. pH may range from 2.5 for saline acid sulfate
soils to 10.5 for soda solonchaks, the texture from sandy to
clay, the organic matter content from 1 to 50%, and the nutrient status from very low to moderately high. Most coastal
saline soils in the humid tropics are subject to flooding by
fresh water during the wet season and submergence by tidal
salt water in the dry season.

Mineral stresses other than salinity are (a) deficiencies
of nitrogen, phosphorus, and zinc, and excess of boron in
arid saline soils, (b) iron toxicity and phosphorus deficiency in acid saline soils, (c) nitrogen, phosphorus, zinc, and
copper deficiencies, and toxicity of organic substance on
coastal peat soils.

Salinity and plant growth. Strongly saline soils are barren of plants. Less strongly saline soils in arid areas

permit growth of halophytic grasses and shrubs (FAO/UNESCO, 1973), whereas coastal saline soils in the humid tropics and subtropics support growth of mangrove species (Chapman, 1976).

Most plants suffer salt injury at EC values exceeding 4 mmho/cm at 25°C (United States Salinity Laboratory Staff, 1954), but Boyko (1966) found that many crop plants can withstand much higher salt concentrations if the soil solution is physiologically balanced as in seawater. Epstein and Norlyn (1977) grew barley in sand dunes irrigated with seawater and obtained a grain yield half as great as the national average. According to Maas and Hoffman (1977), crop yield decreases markedly with increase in salt concentration, but the threshold concentration and rate of yield decrease vary with the species. The effect of salt concentration on yield follows the equation

$$Y = 100 - B (EC_e - A)$$

where Y is relative yield, A is the salinity threshold value, and B is the yield decreases per unit of salinity increase. Panaullah (1980) found that A was higher and B was lower for two salt tolerant rice lines than for two sensitive ones.

FAO/UNESCO (1973) classified crops according to salt tolerance:

Highly tolerant: date palm, barley, sugar beet, cotton, asparagus, spinach;
Moderately tolerant: olive, grape, wheat, oats, rice, maize, alfalfa, potatoes, tomato;
Sensitive: sugar cane, beans, most temperate fruit trees.

A large number of traditional and modern rices that are salt tolerant have been discovered (Ikehashi and Ponnamperuma, 1978).

Reclaiming saline lands. Saline lands can be reclaimed for agricultural production by preventing the influx of salt water, leaching the salts from the root zone, and correcting soil toxicities and nutrient deficiencies. Reclamation costs can be minimized by using salt tolerant species or cultivars.

Alkali or Sodic Soils

Extent and distribution. Alkali or sodic soils cover over 538 million ha of the earth surface (Massoud, 1974). In Sudan, Egypt, Pakistan, and India, great effort is made to bring these soils under food crop production.

Characteristics. Sodic soils contain sufficient exchangeable sodium to depress plant growth, which occurs when exchangeable sodium percentage exceeds 15. pH is usually above
8.5, and the soils contain sodium carbonate or bicarbonate,
calcium carbonate, and high concentrations of water-soluble
silicon. Clay and organic matter are dispersed, so the soils
are sticky when wet and hard when dry. Their internal drainage is poor.

Sodicity and plant growth. Above a total alkalinity of
0.07% HCO_3 and pH 8.7, many crop plants are injured, and
above pH 9.5 nearly all plants die. An exchangeable sodium
percentage of 25 to 30% renders a soil barren and unsuitable
for tillage or irrigation (Kovda, 1965).

The primary growth-limiting factors in sodic soils for
dryland crops are: (a) high pH per se, (b) deficiencies of
available phosphorus, calcium, iron, and zinc, (c) boron
toxicity, (d) poor drainage, and (e) mechanical impedance to
root development. Most of these unfavorable factors are
alleviated when sodic soils are flooded for rice cultivation.

FAO/UNESCO (1973) classified crop plants according to
tolerance for sodicity:

Tolerant: alfalfa, barley, beets (garden and sugar),
bermudagrass, cotton;
  Moderately tolerant: oats, rice, rye, wheat;
  Sensitive: beans, corn, fruit trees.

Alkali tolerant rice lines have been identified and some
are grown on alkali soils (Ikehashi and Ponnamperuma, 1978;
Central Salinity Research Institute, 1979).

Reclaiming sodic soils. Sodic soils can be reclaimed for
agricultural use by applying gypsum and leaching. The gypsum
requirement and the amount of water needed for leaching can
be reduced by using species or cultivars that tolerate sodicity.

Strongly Acid Soils

Nature and extent. Strongly acid soils, which are the
Ultisols, Oxisols, and Spodosols, cover nearly 2.6 billion ha,
mostly in humid regions (Dudal, 1976). Over 2 billion ha of
Oxisols and Ultisols occur in the tropics. Generally, they
are uncultivated but they have potential for food production.

Characteristics. The pH of strongly acid soils is <5.5,
the cation exchange capacity and base status are low, and the
degree of aluminum saturation and the capacity to fix phosphate are high (Dudal, 1980; Kamprath, 1980; Sanchez and
Cochrane, 1980). The adverse effects of low pH are less
severe in Ultisols than in Oxisols (van Wambeke, 1976).

Soil acidity and plant growth. The growth-limiting factors in strongly acid soils for dryland crops are deficiencies of phosphorus, calcium, magnesium, and molybdenum and
toxicities of aluminum and manganese. For wetland rice, the
main problem is iron toxicity (Ponnamperuma, 1978).

There are great inter- and intraspecific differences in
tolerance to soil acidity in tropical soils (Sanchez, 1976;
Spain, 1976). Pineapple, tea, coffee, rice, black beans,
cashew, cassava, cowpeas, peanuts, and several species of
tropical grasses and pasture legumes are well adapted to acid
soil conditions, but maize, sorghum, and cotton are sensitive.
Marked cultivar differences in tolerance for acid soil conditions exist in dryland rice, wheat, sorghum, and cassava
and for iron toxicity in wetland rice.

Amelioration of strongly acid soils. Acid soils can be
made productive by liming and applying nitrogen, phosphorus,
and potash. The amounts of these can be reduced by using
tolerant cultivars.

Acid Sulfate Soils

Extent and distribution. Acid sulfate soils cover about
5 million ha in Southeastern Asia, 4 million in Africa, and
2 million in Venezuela (van Breemen, 1980). Millions of ha
of peats in the tropics are underlain by potential acid sulfate soils.

Characteristics. Acid sulfate soils have a high sulfate
content and an air-dry pH <4.0 in some layer of the upper 50
cm of the profile. They are derived from marine sediments
high in pyrite and poor in bases. When submerged and anaero-
bic, they are nearly neutral and support salt-tolerant marsh
plants, but when drained, the pyrite is oxidized to sulfuric
acid.

Acid sulfate soils and plant growth. The growth-limiting
factors in drained acid sulfate soils are low pH, aluminum
toxicity, high electrolyte content, and low availability of
nitrogen and phosphorus. When flooded, the pH increases, aluminum toxicity decreases, and nutrient availability increases,

but iron toxicity becomes a serious retarding factor (Ponnamperuma et al., 1973). Crops adapted to acid sulfate soils
are pineapple, cashew, and coconut. Rice cultivars with tolerance for acid sulfate soil conditions have been identified
(Ponnamperuma and Solivas, 1981).

   Reclaiming acid sulfate soils. Reclamation measures for
acid sulfate soils include ridging, leaching, and liming.
Tolerance for aluminum and iron toxicities and phosphorus
deficiency in wetland rice will confer advantages.

Peat Soils (Histosols)

   Extent and distribution. Peat soils cover about 240 million ha of the earth's surface (Dudal, 1976). Mostly they
occur in Russia, Europe, and North America (Brady, 1974).
The tropics has 32 million ha (Driessen, 1978).

   Characteristics. Histosols have a surface horizon at
least 50 cm deep which contains 12–18% organic carbon. Unless drained, they are submerged or saturated with water.
Peats may be eutrophic (fertile) but deep or strongly acid
ones are oligotrophic (infertile).

   Histosols and plant growth. The primary factors limiting
plant growth in Histosols are deficiencies of nitrogen, phosphorus, potassium, copper, molybdenum, and zinc. With good
management, temperate peats can be highly productive, but in
the tropics, draining peat soils leads to their rapid oxidation and subsidence. Rapid oxidation can be avoided by
growing wetland rice with reduced micronutrient requirements.
Several such cultivars have been identified (Ikehashi and
Ponnamperuma, 1978; IRRI, 1980).

Phosphorus-Deficient Soils

   Classes. Phosphorus deficiency occurs as a mineral
stress in soils with low total phosphorus content (Kawaguchi
and Kyuma, 1977) and in those that fix phosphorus strongly.
Ultisols, Oxisols, Spodosols, Vertisols, Histosols, Andepts,
and sodic soils are deficient in available phosphorus (Brady,
1974; Dudal, 1976; Sanchez, 1976; van Wambeke, 1976; Dabin,
1980; Sanchez and Cochrane, 1980).

   Correction of phosphorus deficiency. The simplest method
of correcting phosphorus deficiency is to apply this element
as a fertilizer. About 25 million tons of phosphorus

fertilizer is used annually (Dabin, 1980). But with huge increases in the price of phosphate fertilizer, attention is
being given to enhancing the efficiency of phosphorus utilization by crops and to using forms of phosphorus that are
energy-saving to manufacture.

The efficiency of phosphorus fertilizer use is rarely more
than 10% for crops, because of the fixation of this element
by exchangeable aluminum and soil minerals, chiefly halloy-
site, kaolinite, and amorphous clays. van Wambeke (1976) reports that 5000 kg P_2O_5/10 cm/ha of soil was required to
build up the phosphorus availability to the desired level in
Andept.

Some species of tropical legumes have low phosphorus requirements (Kretchmer, 1976), and cassava can extract phosphorus from insoluble phosphorus fertilizers (Cock and
Howeler, 1978). Cultivar differences exist in efficiency of
phosphorus use in rice (Ponnamperuma and Castro, 1972;
Ikehashi and Ponnamperuma, 1978), sorghum (Clark et al.,
1978), and beans (Malavolta and Amaral, 1978).

Iron-Deficient Soils

Iron deficiency is a growth-limiting factor for most dryland crops growing on calcareous and sodic soils (Brown,
1976a). It is the most important nutritional factor limiting
the growth of dryland rice (Ponnamperuma, 1975) and the second
most important micronutrient problem in India (Indian Council
of Agricultural Research, 1978-79). In tropical America 98
million ha of the high base status soils are iron deficient
(Sanchez and Cochrane, 1980). In Australia 34,000 ha are
treated annually with iron chemicals (Donald and Prescott,
1975). Treatments for iron deficiency include soil acidifi-
cation, and applying iron salts or chelates to the soil or
crop. But these methods are uneconomic, disturb nutrient
balance, or cause pollution.

Australian workers, as early as 1957, proposed planting
less susceptible species of tree and field crops (Donald and
Prescott, 1975). Brown et al. (1972) stated, "Iron chlorosis
is one of the most difficult micronutrient deficiencies to
correct in the field."

Citrus and grape growers have controlled iron deficiency
by using iron-efficient rootstocks (Brown, 1976b). Marked
intraspecific differences in tolerance for iron unavailability
have been demonstrated in corn, soybeans, tomato, and rice.
Developing genotypes with tolerance to iron deficiency appears
to be a good solution on calcareous and sodic soils (Brown et
al., 1972).

Zinc-Deficient Soils

Zinc deficiency is an important and widespread disorder of crops the world over. It occurs on sandy, calcareous, sodic, and peat soils. It occurs throughout the United States and several European countries (Lindsay, 1972), on 5.8 million ha of wheat and rice lands in India, 0.7 million ha of rice land in Pakistan, and 2 million ha of irrigated rice lands in the Philippines. In Australia, 374,000 ha of cropland are annually treated with zinc fertilizer (Donald and Prescott, 1975). Sanchez and Cochrane (1980) consider zinc deficiency as a soil constraint on crop productivity in tropical America on 96 million ha of high base soils and 645 million ha of infertile acid soils. Zinc deficiency is a severe growth-limiting factor of wetland rice on sodic, calcareous, organic, and poorly drained soils, and on soils high in water-soluble silicon.

Zinc deficiency is easily corrected by applying zinc compounds to the soil or plant. Soil applications may be as high as 40 kg zinc sulfate per ha per year. Use of zinc efficient varieties will eliminate the need for zinc fertilizers on marginally deficient soils and reduce the amount needed on strongly deficient soils.

Species differ in their zinc-feeding power, and so do varieties within a species (Brown et al., 1972). Strong varietal differences exist in rice (Ponnamperuma and Castro, 1972; Ikehashi and Ponnamperuma, 1978; IRRI, 1980).

Other Nutrient-Deficient Soils

Almost all cropped lands are nitrogen deficient. The use of nearly 50 million tons of nitrogen fertilizer annually proves this point. The demand for nitrogen fertilizer could be reduced by using cultivars that absorb and metabolize soil and fertilizer nitrogen efficiently. Such efficient rice cultivars have been discovered (IRRI, 1980; Ponnamperuma, 1979).

Deficiencies of potassium, sulfur, copper, manganese, and molybdenum are common in many soils, and these deficiencies can be corrected by soil amendments. Cultivar differences in efficiency of use of these elements exist.

With boron, the problem usually is one of toxicity, which cannot be corrected easily by soil amendments. Alfalfa and beets are tolerant; cereals and cotton are moderately tolerant; and citrus and deciduous fruit trees are sensitive (Brown, 1976a). Rice varieties show differential tolerance (IRRI, 1980).

II. VARIETAL TOLERANCE FOR SOIL STRESSES

Nature abounds with plant families or species adapted to
mineral stresses (Epstein, 1972). The existence of variation
among genotypes of plants (Lauchli, 1976) is due to the evolutionary responses of plants to their habitats. Even within
species ecotypes have different adaptations to specific mineral stresses (Epstein, 1972). Although cultivar differences in
tolerance to soil toxicities or nutrient deficiencies were
recognized 50 years ago (Vose, 1963), only during the past
decade has this variation been used in breeding crop plants to
tolerate mineral stresses.

Gregory and Crowther (1928) observed that barley cultivars
responded differentially to nitrogen, phosphorus and potassium fertilizers, and conceived the possibility of breeding
varieties suited to nutrient-deficient soils. The next four
decades saw numerous studies on cultivar differences in
nutrient uptake, fertilizer responsiveness, and soil toxicities (Vose, 1963). The only practical application of cultivar
tolerance for nutrient deficiencies during that period, however, was the use of iron-efficient rootstocks in controlling
iron deficiency in grapes (Brown, 1976a).

Hayward and Wadleigh (1949), reviewing salt and alkali
tolerance studies in the USA, quoted a 1905 publication which
said, ". . . the development of a more alkali-resistant beet
would make it possible to extend considerably the area now
available for beet crops." They reported marked differences
in salt tolerance among genera, species, and cultivars of
agricultural crops.

Salinity Tolerance

Hayward and Wadleigh (1949) reviewed 121 papers dating
from 1886 and noted the value of cultivar tolerance in alfalfa, barley, beet, carrots, cucurbits, lettuce, onions, wheat,
and fruit trees, for crop production on marginal saline lands.
During the period 1940-1975, there were over 600 publications
on salt tolerance in crop plants, of which 30 related to cultivar differences in food crops. Cultivar differences occur
in rice, wheat, barley, coarse grains, tomato, sugar cane,
sugar beet, oil plants, fruit trees, grasses, legumes, and
cotton.

Wetland rice. Breeding for salt tolerance began about 40 years ago and the salt-tolerant cultivar, Pokkali, was recommended for the saline tracts in Sri Lanka (Fernando, 1949). In India salt-tolerant cultivars were crossed with high yielding rices and the selections 47-22 and SR 26 B released in the 1940's (Ikehashi and Ponnamperuma, 1978). More recently tolerant rices have been crossed with the improved modern variety IR8 (Central Soil Salinity Research Institute, 1979). In Egypt the salt-tolerant cultivar Giza 159 is grown in saline tracts (Khalid et al., 1975).

During 1969-1980, The International Rice Research Institute (IRRI) screened 48,671 rice lines for salinity tolerance, including strains from the world germplasm collection, elite breeding lines, and progeny from the salinity-hybridization program. Over 9,000 were tolerant (Table 4.2).

Table 4.2. Summary of screening tests for adverse soils tolerance in rice, 1969-1980.

Stress	Number tested	Number tolerant
Salinity	48,671	9,206
Alkalinity	10,585	2,254
Peat soil problems	1,281	81
Iron toxicity	2,285	85
Phosphorus deficiency	4,178	451
Zinc deficiency	11,600	817
Iron deficiency (dryland rice)	572	68
Aluminum and manganese toxicities (dryland rice)	505	46

Source: Ponnamperuma et al., 1981.

Salt tolerance has been incorporated into rice lines with improved plant type and resistance to disease and insects, and eight such salt-tolerant lines have produced well in farmers' fields in saline tracts in Southern and Southeastern Asian countries. Several improved rice cultivars such as IR42, IR43, and IR52 have salt tolerance, even though no planned effort was made to breed this trait into them (Table 4.3). In coastal saline fields, salt-tolerant rice cultivars yielded on the average 2 t/ha more than non-tolerant ones (Table 4.4).

Table 4.3. Reactions of IR varieties to adverse soil conditions on the scale 1-9.[a]

| | Wetland soils | | | | | | | Dryland soils | |
| | Toxicities | | | | | Deficiencies | | Deficiency | Toxicities |
	Salt	Alkali	Peat	Iron	Boron	Phos-phorus	Zinc	Iron	Aluminum and manganese
IR8	3	6	5	7	4	4	4	4	4
IR26	5	6	6	6	4	1	5	4	3
IR28	7	5	6	4	4	3	5	6	5
IR36	3	3	3	3	3	7	2	2	4
IR42	3	4	3	3	4	3	4	6	5
IR43	3	7	7	4	3	3	3	5	3
IR45	4	7	6	5	3	3	4	4	4
IR52	3	4	3	4	3	3	4	3	5
IR54	4	5	3	3	3	2	2	4	4

[a] 1 = almost normal plant; 9 = almost dead or dead plant.

Table 4.4. Yield advantage due to soil stress tolerance in modern rices as shown by tests in farmers' fields in the Philippines (1977-1981).

Stress	Total number			Mean yield (t/ha)		
	Tests	Sites	Rices	Min.	Max.	Adv.
Salinity	23	14	63	1.5	3.6	2.1
Alkalinity	3	2	47	0.9	3.6	2.7
Iron toxicity	12	4	55	2.2	4.8	2.6
Peatiness	13	5	39	1.4	3.1	1.7
Al/Mn toxicity	3	1	32	2.0	3.8	1.8
P deficiency	13	2	110	1.9	4.4	2.5
Zn deficiency	25	10	91	0.8	2.9	2.1
Fe deficiency	8	3	65	0.9	2.8	1.9

Wheat, barley, coarse cereals. Maliwal and Paliwal (1967) observed that two of six wheat and two of six barley cultivars were highly tolerant of salt and alkali. The barleys were more tolerant than the wheats. Sharma et al. (1971) reported that local Indian wheat cultivars were more tolerant than Mexican strains, and Bains et al. (1971) found that the barley strain IB226 had only a 39% yield reduction compared with 85% for another line when the salt level was raised from a conductivity of 8 to 24 mmho/cm. Three salt-tolerant barley lines irrigated with seawater on a pure sand medium gave nearly half of the average national yield (Epstein and Norlyn, 1977). More recently, Epstein et al. (1980) identified 34 lines of spring wheat that produced seed at a salinity of 50% of that of seawater. Sorghum cultivars differed in their reactions to increasing salt concentrations in solution culture (Taylor et al., 1975).

Tomato. Epstein (1976) found no large differences in salt tolerance among cultivars of tomato. However, by using genes from the wild species, *Lycopersicon cheesmanii*, he obtained progeny that produced acceptable fruits at salinity levels corresponding to 33% that of seawater.

Alkalinity or Sodicity

Work on genetic tolerance for sodicity is meager, and much work reported for dryland crops on alkali soils applies to calcareous soils, but not to sodic or alkali soils. Ikehashi and Ponnamperuma (1978), summarizing the work done on wetland

rice in India prior to 1972, reported two of 12 cultivars
tested on a soil with a pH of 8.6 showed less alkali injury
than the others, and yielded 1.3 and 1.5 t/ha. After tests
in 27 centers in South India with varying degrees of alkalinity, PVR1 was released for production on moderately alkaline
soils. Wheat cultivars with a yield potential of 3 t/ha,
rices with 5 t/ha, and barley cultivars yielding 2.5 t/ha
were released for use on sodic soils in India during the last
decade (Central Soil Salinity Research Institute, 1979).

   In 1972, IRRI found that two out of 66 cultivars tested
showed tolerance to sodicity. By 1980, the number of lines
screened was 10,585 of which 2,254 were tolerant of alkali.
Several improved breeding lines with alkali tolerance are
being tested in Australia, Egypt, India, Mexico, Pakistan,
and Sudan. Those lines have a yield advantage of over 2 t/ha
(Table 4.4).

Aluminum Toxicity

   The most important growth-limiting factor in aerobic
strongly acid soils is aluminum toxicity, so differences
observed in production among crop cultivars on acid aerobic
soils have been attributed to aluminum tolerance (Foy, 1976).
Howeler and Natividad (1976) observed a good correlation between the reactions of dryland rice lines to a high aluminum
concentration in culture solution and performance on a strongly acid Oxisol.

   Wheat. When the acid soils in Brazil were opened to wheat
culture in 1919, only tolerant cultivars gave acceptable
yields; sensitive cultivars perished (Silva, 1976). Tolerance for aluminum toxicity allows wheat production on
acid soils without liming (Silva, 1976). Tolerant cultivars
suffered no injury up to 30% aluminum saturation of the
soil's cation exchange capacity. Tolerance was inherited
complexly as a dominant trait.

   The review of the literature on tolerance of wheat for
aluminum (Reid, 1976) reveals that: (a) there are great cultivar differences in tolerance for aluminum-toxic soils; (b)
older cultivars developed on aluminum-toxic soils have greater aluminum tolerance than those developed on non-toxic
soils; (c) hexaploid wheats have the highest degree of tolerance within the genus *Triticum*; and (d) tolerance for low pH
and aluminum is genetically controlled.

Barley. Marked differences have been observed among barley lines for tolerance for aluminum toxicity. In winter barley this trait is due to a single dominant gene (Reid, 1976).

Rice. Much dryland rice is grown on acid soils in the tropics where aluminum toxicity is a growth-limiting factor (Ponnamperuma, 1975). Tests with 48 cultivars revealed marked differences in growth and yield. Ml-48, a Philippine dryland cultivar, was the top yielder; Peta, a typical wetland rice, was the poorest (Ponnamperuma and Castro, 1972). Howeler and Natividad (1976) observed that traditional rice cultivars from acid dryland areas were more tolerant of aluminum than modern wetland cultivars. Ponnamperuma et al. (1981) found that of 505 rice lines screened, 7 improved selections of wetland rices, 4 traditional dryland, and 1 improved dryland cultivar tolerated acid dryland soils.

Aluminum toxicity restricts root growth, which in turn aggravates drought stress. IR43 and IR45, dryland cultivars from IRRI, have moderate tolerance to aluminum (Ponnamperuma et al., 1981).

Acid Sulfate Soil Problems

Greenhouse and field tests of 190 rice selections for iron toxicity on acid sulfate soils in 1980 showed five were tolerant. Tolerant selections had a yield advantage of 2.6 t/ha over nontolerant lines (Table 4.4). Cultivar tolerance was a suitable substitute for lime on an acid sulfate soil (Table 4.5).

Peat Soil Problems

Rice lines were markedly different in greenhouse and field tests (IRRI, 1980, 1981a) with respect to tolerance for zinc deficiency and other problems on peat soils. Yield advantage due to tolerance was 1.7 t/ha (Table 4.4).

Iron Toxicity

Iron toxicity is a nutritional disorder of wetland rice on strongly acid soils and acid sulfate soils in Colombia, Liberia, Sierra Leone, India, Sri Lanka, Malaysia, Thailand, and Indonesia. It can be alleviated by liming, applying manganese dioxide, draining the soil, and by prolonged submergence, but cultivar tolerance is the simplest solution.

Table 4.5. Varietal tolerance as a substitute for soil amendments on some problem wetland rice soils.

Grain yield (t/ha)

Acid sulfate soil	No lime	5 t lime/ha	P-deficient soil	No P	20 kg p/ha	Zn-deficient soil	No Zn	ZnO dip
IR43	6.2	6.2	IR54	4.7	5.1	IR54	4.4	5.2
IR26	3.9	4.8	IR36	2.4	4.1	IR26	3.1	4.2

Screening of rice lines in Sri Lanka, Liberia, and the Philippines has shown marked strain differences in tolerance to iron toxicity (Gunawardena, 1975; Ikehashi and Ponnamperuma, 1978; Virmani, 1978). One cultivar released as a variety for growing in the acid swamps in Liberia has good tolerance (Virmani, 1978).

Screening for tolerance of rice to iron toxicity began at IRRI in 1970, and to date, of 2,285 lines tested, 85 were tolerant. Among them were traditional cultivars from iron-toxic areas in Asia and several IR lines with good agronomic characteristics and resistance to diseases and insects.

Iron Deficiency

Iron deficiency is one nutritional disorder that cannot be corrected economically by applying chemicals. Cultivar tolerance may be the only solution for dryland crops.

Brown and his colleagues have demonstrated genotypic differences in efficiency of iron utilization in corn, tomato, soybean, and sorghum (Brown et al., 1972; Brown, 1976a, 1976b; Brown and Jones, 1977; Brown, 1978; Brown, 1981). Of 317 rice lines tested for tolerance to iron deficiency at IRRI, 34 were tolerant (Ponnamperuma, 1974, 1975). In rice, tolerance confers a yield advantage of about 2 t/ha (Table 4.4).

Phosphorus Deficiency

Genotypic differences for response to phosphorus deficiency have been reported in beans, soybean, rice, sorghum, and white clover (Ponnamperuma and Castro, 1972; Gerloff, 1976; Cardus and Dunlop, 1978; Clark et al., 1978; Gunawardena and Wijeratne, 1978; Malavolta and Amaral, 1978). In wetland rice, of 4,178 lines screened, 451 were efficient in phosphorus use. Among them are the modern, disease- and insect-resistant cultivars IR50, IR52, and IR54. Phosphorus deficiency tolerance confers a yield advantage of over 2 t/ha (Table 4.4) and is a substitute for phosphorus fertilizer (Table 4.5).

Zinc Deficiency

Zinc deficiency is a widespread disorder of wetland rice, so IRRI has screened nearly 12,000 rice lines on zinc-deficient soils. Of these, 817 were tolerant, and among these were several modern, disease- and insect-resistant cultivars. Use of tolerant improved cultivars on zinc-

deficient soils confers a yield advantage of over 2 t/ha over
comparable nontolerant cultivars (Table 4.4). Tolerance may
be a substitute for zinc fertilizer (Table 4.5).

Mechanisms of Tolerance

Genotypic differences in tolerance to soil toxicities and
mineral deficiencies may be due to genetic variations in ion
absorption by roots, translocation of the elements into and
through the xylem, retention of the ions in adjoining tissues,
mobility in the phloem, and efficiency of metabolic ion utilization (Epstein, 1972). Lauchli (1976) emphasized that genetic differences in ion uptake or transport is important in
breeding for salt tolerance.

III. BREEDING RICE FOR SOIL STRESSES

Recent research at IRRI and other research centers has
shown that: (a) substantial differences exist among rice
lines for tolerance to soil mineral stresses; (b) tolerance
for salinity, aluminum toxicity, and iron deficiency are
genetically controlled; (c) the mode of gene action has been
established for some cases of tolerance to soil stresses; (d)
techniques for screening for mineral stress tolerance have
been developed and used routinely; (e) breeding procedures
have been developed to enhance tolerant levels; (f) economically significant improvements in rice performance have been
achieved; and (g) the prospect of developing rice lines
suited to most adverse soil conditions is good. The dramatic
success of the Genetic Evaluation and Utilization (GEU) program at IRRI illustrates the value of the genetic approach to
solving stress problems.

Genetic Evaluation and Utilization of Rice (GEU)

The introduction of IR8 rice cultivar in 1966 ushered in
the green revolution for rice, and surplus rice production
was predicted for the seventies. But, in 1977 the International Food Policy Research Institute (1977) warned that by
1990 there would be a severe rice shortage. Rice shortages
would occur because the green revolution bypassed the small
farmers who produce the bulk of rice in Southern and Southeastern Asia. Because small farmers could not afford to provide the water control, soil amendments, and pesticides to
exploit the high-yielding cultivars, it became obvious that

rice breeders must tailor the cultivar to fit the production
environment of the small farmers. GEU was launched in 1973
(IRRI, 1974) to build into rice cultivars tolerance to adverse
factors of the environment of small farmers. The adverse components were: (a) too little or too much water; (b) diseases;
(c) insect damage; (d) deficiencies of nitrogen, phosphorus,
zinc, and iron; and (e) soil toxicities such as salinity,
alkalinity, strong acidity, iron toxicity, and excess organic
matter. The "adverse-soils" component of the GEU program was
a combined effort by plant breeders and soil chemists to select and breed rices for soil mineral stresses.

Screening. First, the world collection of rice was evaluated for lines that possessed tolerances to the different
soil stresses. The tolerant entries from the world collection
were used in a hybridization program with good agronomic types
and elite lines from this breeding program were screened. A
summary of these tests is given in Table 4.2.
    Greenhouse and field techniques suited to each stress have
been developed (Ponnamperuma, 1976), and promising selections
are tested at many locations in farmers' fields in the Philippines and in the International Rice Testing Program. Finally,
replicated yield trials are conducted in farmers' fields where
specific soil stresses occur.

Breeding. The primary effort at IRRI has been to breed for
salt and alkali tolerance. Generally, tolerant lines from the
rice world collection were tall cultivars with low yield
potential and little disease and insect resistance. By
crossing them with the modern, disease- and insect-resistant
genotypes, segregates were developed that combined high yield
potential and salt or alkali tolerance. Outstanding selections are IR4630-22-2 and IR9884-54-3 for salinity and IR9715-
7-2 for alkalinity tolerance.
    Our breeding program utilizes: (a) population breeding
methods to accumulate genes from diverse sources; (b) incorporation of multiple soil stress tolerance into a single
breeding line; (c) selecting for high yield and biological
stress tolerance from among the tolerant lines; (d) breeding
for a specific soil stress that may be a major constraint on
productivity; (e) cell and tissue culture in combination with
mutagenesis; (f) exploring exotic germplasm to enhance tolerance; (g) exploiting hybrid vigor; and (h) studying the physiology and genetics of tolerance to soil stresses.

Mineral stress tolerance in elite breeding lines. Many
elite breeding lines have a degree of stress tolerance comparable with that of tolerant donor genotypes. In 1979, of
the 25,086 rice lines screened, 4,532 were found tolerant of
adverse soil conditions, and 3,227 were selections from the
general breeding program (IRRI, 1980). Because these elite
lines have high yield potential and resistance to diseases
and insects, some are gaining acceptance in Southern and
Southeastern Asia even though they are not released as cultivars.

Multiple stress tolerance. Some rice lines developed at
IRRI have multiple tolerances to mineral stress combined with
good agronomic characteristics and resistance to insects and
diseases (Table 4.3). Probably, the multiple stress tolerance
of IR36 and IR42 was a reason for their wide acceptance in
Southern and Southeast Asia (IRRI, 1981b). IR52 and IR54, two
recent cultivars with multiple mineral stress tolerance, were
among the best yielders in the 1980 international tests.

Varietal tolerance and yield stability. Mahadevappa et
al. (1979) observed that tolerance to mineral stresses in
rice varied widely with the stress level and the genotype.
Sensitive rices suffered severe yield losses even under mild
stress, whereas tolerant rices resisted the yield decline
until the stress became moderate. Under severe stress both
tolerant and sensitive genotypes died.
   Genetic tolerance contributed 1.7 to 2.7 t/ha, depending
on the stress, at yield levels of 3 to 5 t/ha. So, these
tolerant rice lines produced economically satisfactory yields
under mineral stresses common in rice land without soil amendments.

International rice testing program. The International Rice
Testing Program is a part of the GEU program which enables the
evaluation of rice lines for tolerances to soil under the exact conditions in which the mineral stresses occur. The results of these tests are published and seed of promising lines
is made available to interested researchers. Currently tests
are conducted in 63 countries for the following soil stresses:
salinity, alkalinity, peat soil conditions, acid sulfate soil
conditions, iron toxicity, and acid dryland conditions.

SUMMARY

The pressing need for more food, the scarcity of arable land, and the high cost of fertilizers focus attention on breeding crop plants adapted to toxic and nutrient-deficient soils. Marked genotypic differences exist in crop plants in tolerance for salinity, alkalinity, strong acidity, and nutrient deficiencies that are not easily corrected by soil amendments. Techniques have been developed for mass screening of cultivars for soil stresses. Rice breeders have combined tolerance for soil stresses with improved agronomic character- istics and resistance to insects and diseases. To produce cultivars have given economic returns on marginal lands and on nutrient-deficient soils without costly inputs. Tolerance for a soil stress can confer a yield advantage of about 2 t/ha, and such rice cultivars with multiple stress tolerances have gained wide acceptance in Southern and Southeastern Asia.

REFERENCES

Akbar, M., and Ponnamperuma, F. N. (1980). "Rice Strategies for the Future." Int. Rice Res. Inst., Los Baños, Philip- pines.

Bains, S. S., Singh, K. N., and Bakshi, D. (1971). *Indian J. Agron. 14*, 356.

Boyko, H. (1966). *In* "Salinity and Aridity" (H. Boyko, ed.), p. 131. Dr. W. Junk Publishers, The Hague.

Brady, N. C. (1974). "Nature and Properties of Soils" 8th ed., p. 1. Macmillan, New York.

Brown, J. C. (1976a). *Proc. Workshop on Plant Adaptation to Mineral Stress in Problem Soils*, p. 83. Beltsville, Mary- land. Nov. 22-23, 1976.

Brown, J. C. (1976b). *Proc. Workshop on Plant Adaptation to Mineral Stress in Problem Soils*, p. 355. Beltsville, Maryland. Nov. 22-23, 1976.

Brown, J. C. (1978). *In* "Crop Tolerance to Suboptimal Land Conditions," p. 257. Amer. Soc. Agron. Spec. Publ. No. 32. Madison, Wisconsin.

Brown, J. C. (1981). *J. Plant Nutr. 3*, 523.

Brown, J. C., and Jones, W. E. (1977). *Agron. J. 69*, 410.

Brown, J. C., Ambler, J. E., Chaney, R. L., and Foy, C. D. (1972). *In* "Micronutrients in Agriculture" (J. J. Mort- vedt, P. M. Giordano, W. L. Lindsay, eds.), p. 389. Soil Sci. Soc. Amer. Madison, Wisconsin.

Cardus, J. R., and Dunlop, J. (1978). *Proc. Colloq. on Plant Analysis and Fertilizer Problems*, p. 75. Auckland, New Zealand. Aug. 29-Sept. 1, 1978.

Central Soil Salinity Res. Inst., Karnal. (1979). "A Decade
    of Research," p. 1. Everett Press, Delhi.
Chapman, V. J. (1976). "Mangrove Vegetation," p. 1. J.
    Cramer.
Clark, R. B., Maranville, J. W., and Gorz, H. J. (1978).
    *Proc. 8th Int. Colloq. on Plant Analysis and Fertilizer
    Problems*, p. 93. Auckland, New Zealand. Aug. 29–Sept. 1,
    1978.
Cock, J. H., and Howeler, R. H. (1978). *In* "Crop Tolerance
    for Suboptimal Conditions," p. 145. Amer. Soc. Agron.
    Spec. Publ. No. 32. Madison, Wisconsin.
Dabin, B. (1980). *In* "Soil-related Constraints to Food
    Production in the Tropics," p. 217. Int. Rice Res. Inst.,
    Los Baños, Philippines.
Donald, C. M., and Prescott, J. A. (1975). *In* "Trace Ele-
    ments in Soil-Plant-Animal Systems" (D. J. D. Nicholas and
    A. R. Egan, eds.), p. 7. Academic Press, New York.
Driessen, P. M. (1978). *In* "Soils and Rice," p. 763. Int.
    Rice Res. Inst., Los Baños, Philippines.
Dudal, R. (1976). *Proc. Workshop on Plant Adaptation to
    Mineral Stress in Problem Soils*, p. 3. Beltsville, Mary-
    land. Nov. 22–23, 1976.
Dudal, R. (1978). *Proc. 11th Int. Soil Sci. Cong.* Edmonton,
    Alberta, Canada. June 12–27, 1978.
Dudal, R. (1980). *In* "Soil-related Constraints to Food Pro-
    duction in the Tropics," p. 23. Int. Rice Res. Inst.,
    Los Baños, Philippines.
Epstein, E. (1972). *In* "Mineral Nutrition of Plants: Prin-
    ciples and Perspectives," p. 412. John Wiley and Sons,
    Inc. New York.
Epstein, E. (1976). *Proc. Workshop on Plant Adaptation to
    Mineral Stress in Problem Soils*, p. 73. Beltsville,
    Maryland. Nov. 22–23, 1976.
Epstein E., and Norlyn, J. D. (1977). *Science 197*, 249.
Epstein E., Norlyn, J. D., Rush, D. W., Kingsbury, R. W.,
    Kelly, D. B., Cunningham, G. A., and Wrong, A. F. (1980).
    Science 21, 399.
FAO (Food and Agriculture Organization). (1974). FAO–UNESCO
    soil map of the world, 1:500,000. Legend. UNESCO, Paris.
FAO–UNESCO (Food and Agriculture Organization–United Nations
    Educational, Scientific, and Cultural Organization).
    (1973). "Irrigation, Drainage, and Salinity. An Inter-
    national Source Book." Hutchinson, London. 510 p.
Foy, C. D. (1976). *Proc. Workshop on Plant Adaptation to
    Mineral Stress in Problem Soils*, p. 255. Beltsville,
    Maryland. Nov. 22–23, 1976.

Gerloff, G. C.   (1976).   *Proc. Workshop on Plant Adaptation to Mineral Stress in Problem Soils*, p. 161.   Beltsville, Maryland.   Nov. 22–23, 1976.

Gregory, F. G., and Crowther, F.   (1928).   *Ann. Bot. XLII*, 757.

Gunewardena, I.   (1975).   Int. Rice Res. Conf., Los Baños, Philippines.   Apr. 21–24, 1975.

Gunewardena, S. D. I. E., and Wijeratne, H. M. S.   (1978).   Int. Rice Res. Conf., Los Baños, Philippines.   Apr. 17–21, 1978.

Hayward, H. E., and Wadleigh, C. H.   (1949).   *Adv. Agron. 1*, 1.

Howeler, R. H., and Cadavid, L. F.   (1976).   *Agron. J. 68(4)*, 551.

International Food Policy Research Institute.   (1977).   *In* "Food Needs of Developing Countries: Projections of Production and Consumption to 1990," p. 157.

Ikehashi, H., and Ponnamperuma, F. N.   (1978).   *In* "Soils and Rice," p. 801.   Int. Rice Res. Inst., Los Baños, Philippines.

Indian Council of Agricultural Research 1978–79.   All India Co-ordinated Scheme of Micronutrients in Soils and Plants.   12th Annu. Rep., New Delhi.

IRRI.   (1974).   Annu. Rep. 1973.   Los Baños, Philippines.   266 p.

IRRI.   (1980).   Annu. Rep. 1979.   Los Baños, Philippines.   538 p.

IRRI.   (1981a).   Annu. Rep. 1980.   Los Baños, Philippines.   (in press).

IRRI.   (1981b).   "Research Highlights for 1980."   Int. Rice Res. Inst., Los Baños, Philippines.

Kamprath, E.   (1980).   *In* "Soil-related Constraints to Food Production in the Tropics," p. 171.   Int. Rice Res. Inst., Los Baños, Philippines.

Kawaguchi, K., and Kyuma, K.   (1977).   "Paddy Soils in Tropical Asia."   Univ. Press of Hawaii, Honolulu.   258 p.

Khalil, M. M., Ata, S. K., Sheta, T. H., and Bakkati, H.   (1975).   *Agric. Res. Rev. 53(4)*, 51.

Kovda, V. A.   (1965).   *Agrokemia es Talajtan Tom 14 Supplementum*, 15.

Kretchmer, A. E., Jr.   (1978).   *In* "Crop Tolerance to Suboptimal Land Conditions," p. 97.   Amer. Soc. Agron. Spec. Publ. No. 32.   Madison, Wisconsin.

Lauchli, A.   (1976).   *In* "Encyclopedia of Plant Physiology" (U. Lüttage and M. G. Pitman, eds.), p. 372.   New Series, *2*, Part B.   Springer-Verlag, Berlin.

Lindsay, W. L.   (1972).   *Adv. Agron. 24*, 147.

Maas, E. V., and Hoffman, G. J. (1977). *J. Irrigation and Drainage Div.* ASCE, *103*, No. IR2, Proc. Paper 12993.

Mahadevappa, M. M., Ikehashi, H., and Ponnamperuma, F. N. (1979). Int. Rice. Res. Inst. Res. Pap. Ser. 43. 15 p. Los Baños, Philippines.

Malavolta, E., and Amaral, E. A. L. (1978). *Proc. 8th Int. Colloq. on Plant Analysis and Fertilizer Problems*, p. 313. Auckland, New Zealand. Aug. 29–Sept. 1, 1978.

Maliwal, G. L., and Paliwal, K. V. (1967). *Indian J. Pl. Physiol. 10*, 26.

Massoud, F. I. (1974). "Salinity and Alkalinity as Soil Degradation Hazards." FAO/UNDP Expert Consultation on Soil Degradation, FAO, Rome. June 10–14, 1974. 21 p.

Northcote, K. H., and Skene, J. K. M. (1972). Commonwealth Sci. Indust. Res. Org., *Soil Publ. 27*, 62. Australia.

Panaullah, G. M. (1980). M.S. thesis. Univ. Philippines, Los Baños.

Ponnamperuma, F. N. (1974). *In* "Soil Management in Tropical America" (E. Bornemiza and A. Alvarado, eds.), p. 330. Soil Sci. Dept., North Carolina State Univ., Raleigh.

Ponnamperuma, F. N. (1975). *In* "Major Research in Upland Rice," p. 40. Int. Rice Res. Inst., Los Baños, Philippines.

Ponnamperuma, F. N. (1976). *Proc. Workshop on Plant Adaptation to Mineral Stress in Problem Soils*, p. 341. Beltsville, Maryland. Nov. 22–23, 1976.

Ponnamperuma, F. N. (1978). *Proc. 8th Int. Colloq. on Plant Analysis and Fertilizer Problems*, p. 383. Auckland, New Zealand. Aug. 29–Sept. 1, 1978.

Ponnamperuma, F. N. (1979). Int. Rice Res. Inst. Res. Pap. Ser. 44. Los Baños, Philippines.

Ponnamperuma, F. N., and Bandyopadhya, A. K. (1980). *In* "Soil-related Constraints to Food Production in the Tropics," p. 203. Int. Rice Res. Inst., Los Baños, Philippines.

Ponnamperuma, F. N., and Castro, R. U. (1972). *In* "Rice Breeding," p. 677. Int. Rice Res. Inst., Los Baños, Philippines.

Ponnamperuma, F. N., and Solivas, J. L. (1981). *Proc. 2nd Int. Symp. on Acid Sulfate Soils*. Bangkok, Thailand. Jan. 18–25, 1981. (in press).

Ponnamperuma, F. N., Attanandana, T., and Beye, G. (1973). *Proc. 1st Int. Symp. on Acid Sulfate Soils, (2)*, p. 391. Wageningen. Aug. 13–20, 1972.

Ponnamperuma, F. N., Gunawardena, I., Ikehashi, H., and Coffman, W. R. (1981). "Adverse Soils Tolerance. Internal Program Review." Int. Rice Res. Inst., Los Baños, Philippines. Jan. 19–31, 1981.

Reid, D. A. (1976). *Proc. Workshop on Plant Adaptation to Mineral Stress in Problem Soils,* p. 55. Beltsville, Maryland. Nov. 22-23, 1976.

Sanchez, P. A. (1976). "Properties and Management of Tropical Soils." John Wiley and Sons, New York. 618 p.

Sanchez, P. A., and Cochrane, T. T. (1980). *In* "Soil-related Constraints to Food Production in the Tropics," p. 107. Int. Rice Res. Inst., Los Baños, Philippines.

Sharma, O. P., Puntankar, S. S., and Seth, S. P. (1971). *Indian J. Agric. Sci. 24(11),* 44.

Silva, A. R. da. (1976). *Proc. Workshop on Plant Adaptation to Mineral Stress in Problem Soils,* p. 223. Beltsville, Maryland. Nov. 22-23, 1976.

Soil Science Society of America. (1978). "Glossary of Soil Science Terms." Madison, Wisconsin.

Spain, J. M. (1976). *Proc. Workshop on Plant Adaptation to Mineral Stress in Problem Soils,* p. 213. Beltsville, Maryland. Nov. 22-23, 1976.

Taylor, R. M., Yony, E. F., Jr., and Rivera, R. C. (1975). *Crop Sci. 15,* 734.

United States Salinity Laboratory Staff. (1954). Agric. Handb. 60, 160. USDA, Washington, D. C.

van Breemen, N. (1980). *In* "Soil-related Constraints to Food Production in the Tropics," p. 189. Int. Rice Res. Inst., Los Baños, Philippines.

van Wambeke, A. (1976). *Proc. Workshop on Plant Adaptation to Mineral Stress in Problem Soils,* p. 15. Beltsville, Maryland. Nov. 22-23, 1976.

Vose, P. B. (1963). *Herb. Abstr. 33(1),* 1.

Virmani, S. S. (1978). "Some Results of Varietal Screening for Tolerance to Iron Toxicity." Int. Rice Res. Conf., Los Baños, Philippines. Apr. 17-21, 1978.

Willett, J. W. (1976). "The World Food Situation: Problems and Prospects to 1985." Oceana Publ., Inc., New York. 1136 p.

CHAPTER 5

COMBINING GENOMES
BY CONVENTIONAL MEANS

Jacqueline James Sprague

Centro International de Mejoramiento
de Maiz y Trigo
El Batan, Mexico

Alien genome combination has received considerable attention from several disciplines over the years. Its practical application in the improvement of cultivated plants increases as breeders continue to need specific and general variability from genera outside the range of normally compatible taxa. For such genome combination to be relevant to plant breeding it is necessary to produce viable hybrids and subsequent progenies capable of alien gene transfer.

This paper discusses the subject with particular reference to the cereal crops, and addresses the following topics in relation to the production of a usable hybrid and its subsequent potential for plant improvement:

  I.   HYBRID PRODUCTION VIA THE SEXUAL ROUTE
  II. HYBRID VIABILITY
  III. CHROMOSOME VARIATION IN F_1 PROGENY
  IV. POTENTIAL ALIEN GENOME UTILIZATION

I. HYBRID PRODUCTION VIA THE SEXUAL ROUTE

In general, the more distantly related the parents, the more difficult it is to produce a hybrid between them, because of various hybridization barriers established during the course of evolution. The stages at which these barriers may occur in the conventional, sexual process of hybridization are: (a) Pollen germination, (b) Pollen tube growth, (c) Fertilization, and (d) Seed development.

Pollen germination and pollen tube growth must occur for timely delivery of the male gamete to allow successful fertilization, and endosperm and embryo development must follow to produce viable hybrids capable of gene transfer.

Pollen Germination

Nongermination of pollen has not been a problem for obtaining interspecific hybrids in grasses. For example, sorghum (*Sorghum bicolor*) pollen germinates readily on all regions of the gynoecium: stigma, style, and ovary of both maize (*Zea mays*) and pearl millet (*Pennesetum americanum*) (Reger and James, 1982) (Figure 5.1). Similarly there is no inhibition of pollen germination in wheat (*Triticum aestivum*) x rye (*Secale cereale*) crosses (Lange and Wojciechowska, 1976), nor in wheat x barley (*Hordeum vulgare*) crosses (Snape et al., 1980).

Pollen Tube Growth

Pollen tube growth, on the other hand, has been shown to restrict fertilization between several incompatible grass species.

Our investigations at CIMMYT have shown that restricted pollen tube growth is a major problem in all intergeneric crosses attempted with maize (Figure 5.2). Table 5.1 shows that pollen tubes are observed at the maize micropyle in a very low percentage of gynoecia after alien pollination.

One tissue in which the crossability (Kr_1 Kr_2) genes operate in wheat x rye crosses is in the style (Lange and Wojciechowska, 1976). Similarly, Snape et al. (1980) report that tetraploid *H. bulbosum* pollen tubes do not penetrate the wheat embryo sac in some wheat x *H. bulbosum* crosses.

Figure 5.1.   Sorghum pollen germination on maize ovary.

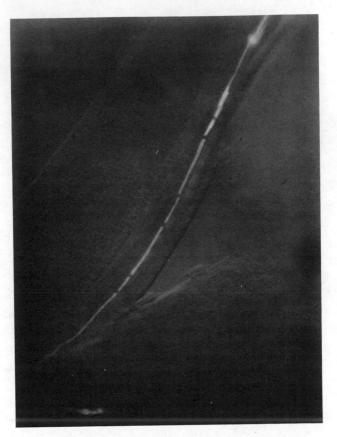

Figure 5.2.   Sorghum pollen tube in maize ovary.

Table 5.1.   Alien Pollen Tube Growth in Maize
            (Amarillo Bajio).

| | | No. pollen tubes | |
| | Number of | in | at micro- |
Pollen parent	observations	ovary	pyle
Tripsacum dactyloides 7138-2	1052	4	3
Tripsacum dactyloides 7153-4	1067	4	2
Sorghum bicolor	3100	6	3

Ionizing radiation may have a practical application in overcoming such hybridization barriers (Pandey, 1974). X-rays can speed up pollen germination, and increase the rate of pollen tube elongation in some crops (Livingston and Stettler, 1973). Tuzuka (1952) obtained hybrids in otherwise incompatible crosses in *Avena* by treating the pollen with X-rays. However in general such techniques have failed to overcome intergeneric incompatibility (Davies and Wall, 1961).

It appears that in some intergeneric crosses in the cereals, sufficient genotypic and/or environmental variation exists for successful pollen tube growth and for timely delivery of the male gamete to the egg cell, to permit subsequent fertilization and genome combination (Zeven and Heemert, 1970; Zenkteller and Straub, 1979). However, in general unsuccessful pollen tube growth is a major barrier to intergeneric hybrid production in the cereals (Snape et al., 1980; Reger and James, 1982), and in many other crops (Linskens, 1972). Consequently considerably more research effort is needed to overcome this barrier. Hybrid production will continue to be restricted until this problem is better understood and the barrier overcome.

Fertilization

The main reason for lack of fertilization after intergeneric pollination in the grasses is nondelivery of the male gametes. In most crosses of sugar cane (*Saccharum officinarum*) and maize for example, it has been reported that delivery of the male gametes results in successful fertilization of the egg and polar cells (Vijendra Das, 1970). However, in some of these crosses the gametes were observed to maintain separate identities in the embryo sac. In other crosses the gametes were reported to fuse, but the proembryo remained in a binucleate condition, the nuclei failing to enter vegetative division. Normal pollen tube growth but nonfertilization of ovules has been observed also for some wheat x rye crosses (Tozu, 1966). The fact that all prefertilization events are reported to be essentially the same in both successful and unsuccessful crosses in such cases, suggests that the lack of fertilization is due to a genetic and/or environmental incompatibility between the individual gametes. However this does not appear to be common, and the problem can be avoided in crosses such as sugar cane x maize (Vijendra Das, 1970) by using the abundant natural variation that exists within the species.

All prefertilization events are influenced by gene action
and environment (Linskens, 1972). In addition to these factors, postfertilization events are influenced by genome compatibility (Gupta, 1969; Davies, 1974). All these factors
may be confounded by the ploidy levels of the parents
(Stebbins, 1950).

Seed Development

Endosperm and embryo breakdown are common after interspecific crosses. In different crops, this has been attributed
to the genetic constitution of the maternal tissue and the
embryo, or of the endosperm and the embryo (Brink and Cooper,
1947), or to the incompatibility of the different ploidy
levels (Stebbins, 1950).
Despite a high frequency of seed collapse, this incompatibility mechanism often is not complete, and some viable embryos develop and can be harvested from shrivelled mature
seeds. For example, this occurs commonly in crosses of maize
x *Tripsacum* spp. and of wheat x rye. In other crosses between more distantly related genera, e.g., maize x sorghum,
no embryo has been reported to have survived naturally to
maturity. Breakdown normally occurs at early stages of seed
development (James, unpubl.). Figure 5.3 shows an ear of
maize 12 days after pollination with sorghum and sections of
selected seed from the same cross, also sampled 12 days after
pollination. Approximately 90% of all collapsed seed possesses no visible sign of embryo nor endosperm development.
The remaining 10% shows some endosperm development, 30% of
which possesses a recognizable embryo. Figure 5.3 (b) shows
an enlarged embryo sac with endosperm at the stage of transition from a multinucleate to a cellular condition, which normally occurs in maize approximately four days after pollination (Kiesselbach, 1949). Figure 5.3 (c) shows the enlarged
cells of an undividing embryo with breakdown endosperm tissue around the edge of the embryo sac. The embryo possesses
no more than 16 cells, which occurs normally approximately
five days after pollination (Kiesselbach, 1949).
Several techniques have been used successfully to overcome such seed breakdown. Some of these bypass or partially
bypass the conventional sexual route to hybridization (Kao,
1975; Ohyama, 1975). Fertilization of cultured ovaries or
ovules can allow embryo development in vitro that may not be
possible in vivo (Dhaliwal and King, 1978). Recent improvements in embryo culture techniques have increased significantly the percentage of viable hybrid embryos that can be obtained from intergeneric crosses.

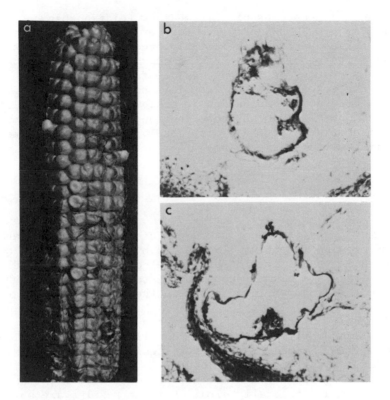

Figure 3.   Seed breakdown in maize 12 days after pollination with sorghum: (a) seed collapse
on ear; (b) enlarged embryo sac with endosperm in transition stage from multinucleate to cellular condition; (c) enlarged cells of non-dividing embryo with
breakdown endosperm tissue around edge of embryo sac.

    Also certain chemical treatments assist seed development.
Pod abortion was delayed in the interspecific cross *Vigna
radiata* x *V. umbellata*, after the application of E-amino n-
Caproic acid (EACA), which allowed a longer period for natural
embryo development on the plant (Baker et al., 1975). Twice
as many seeds developed as compared with non-use of the
chemical. Also with the use of EACA, Taira and Larter (1977)
obtained a better development of hybrid embryos of *Triticale*
than when no EACA was used. Kruse (1973), Islam et al.
(1975), Fedak (1977), Chapman and Miller (1978), and Mujeeb-
Kazi and Rodriguez (1981), routinely use gibberellic acid to
assist in the production of intergeneric hybrids with *Triticum*.
    Since all barriers to sexual hybridization are affected by
evolutionary relationships (Kimber and Sallee, 1976), bridge

crosses with species of intermediary relationship can be
used for hybrid production. However, as more and more hybridization barriers are being overcome, and because of
associated problems in specific character transfer from the
desired species without any associated effects from the
bridge species, this approach is used less frequently today.

Alien genome combination is achieved through fertilization. However, for any contribution to plant improvement it
is imperative that the zygote develop to a viable plant able
to transmit alien germplasm.

II. HYBRID VIABILITY

Many progenies from intergeneric crosses are weak and
slow growing (Stebbins, 1958), and hybrid inviability often
prevents any utilization of genome combinations. Seedling
lethality is a common phenomenon after intergeneric hybridization, and is mentioned often in the literature (Capari,
1948; Stebbins, 1950, 1958).

In some intergeneric hybrids, the expression of lethality
is delayed until plant growth is well advanced. Siddiqui and
Jones (1969) found that some trispecific hybrids of *Triticum
aestivum* with the amphidiploid of *T. durum* x *Aegilops squar-
rosa* lived for over two months, produced tillers, then became
chlorotic and gradually died. Similar results are obtained
with some maize x *Tripsacum* hybrids (James, unpubl.).

Frequently plants from intergeneric crosses die before
their true parentage can be established (Gerrish, 1967).
Such inviability is not necessarily an indication of hybrid-
ity since alien pollination can stimulate apomictic development of a reduced or an unreduced egg cell to give haploid or
maternal progeny respectively (Gustafsson, 1946). Development of a post meiotic cell with or without chromosome
doubling would yield plants that were homozygous or hemizygous respectively for recessive genes. Apomixis, therefore,
could give rise to inviable seedlings because they are homozygous or hemizygous for lethal recessive genes. However,
haploid and maternal progeny are produced also by alien
chromosomal elimination after successful fertilization
(Subrahmanyam and Kasha, 1973; Davies, 1974; Barclay, 1975).

III.   CHROMOSOME VARIATION IN F$_1$ PROGENY

    Somatic and meiotic variations have been reported in
several intergeneric hybrids (Islam et al., 1978; Mujeeb-
Kazi and Rodriguez, 1981b), and can hinder correct classification of the progeny and influence utilization of the alien
genomic combination.

Somatic

    The cross *H. vulgare* x *H. bulbosum* was reported to give
haploids in 1958 (Davies), but the mechanism that gives
these haploids, selective chromosome elimination after fertilization, and not a failure of alien genome combination,
was first elucidated by Kasha and Kao (1970).
    In this cross if both parents are either diploid or tetra-
ploid, all *H. bulbosum* chromosomes are gradually eliminated
in cell divisions at early stages of embryo development,
irrespective of which species is used as the female parent.
On the other hand, if the *H. vulgare* parent is diploid, and
the *H. bulbosum* is tetraploid, no chromosome elimination
occurs, and a triploid hybrid is produced. Fertilization of
T. aestivum with diploid or tetraploid *H. bulbosum* gives the
same result. Gradual elimination of the *H. bulbosum* chromosomes from the embryo of the hybrid occurs to give polyhap-
loid wheat plants (Barclay, 1975).
    Chromosome elimination is not always complete in inter-
generic hybrids. In maize x *Tripsacum* crosses two types of
hybrids are produced (James, 1979). The fertilization of a
normal reduced female maize gamete gives a classical hybrid
with the expected gametic number of chromosomes from both
parents (Figure 5.4). However the fertilization of an unreduced maize gamete has given a gradual and erratic elimination of the alien *Tripsacum* chromosomes throughout the life
of the plant (James, 1979). Figure 5.5 shows a cell from
such a hybrid with 20 maize and one *Tripsacum* chromosome.
This latter phenomenon has been observed also in progenies
from maize x sorghum crosses. For example cells of one hybrid were observed to possess 30 chromosomes, 20 maize and 10
smaller sorghum chromosomes, at early stages of growth, but
as the plant aged, progressively fewer cells of the plant
possessed alien chromosomes in successive samples (James,
1978) (Figure 5.6).
    Chromosome elimination is not limited to alien chromo-
somes. In some maize x *Tripsacum* hybrids (James, 1979), some
cells possessed 20 chromosomes up to four of which were

Figure 5.4.   Classical maize x Tripsacum hybrid:
(a) phenotype; (b) cytology, 10 maize and 36 Tripsacum
chromosomes.   Two detached satelite segments of maize
chromosome arrowed.

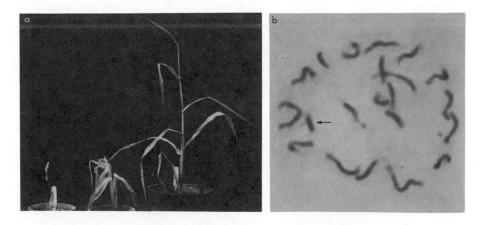

Figure 5.5.   Non-classical maize x Tripsacum hybrids:
(a) phenotype; (b) cytology, 20 maize and 1 Tripsacum
(arrowed) chromosome.

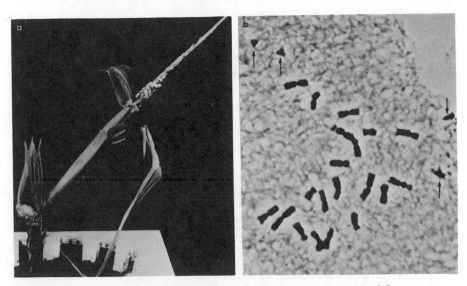

Figure 5.6. Non-classical maize x sorghum hybrid:
(a) phenotype; (b) cytology, 20 maize and 4 sorghum (arrowed)
chromosomes.

Tripsacum. Also mitotic disturbances during early zygotic
divisions of wheat x barley hybrids cause elimination and
duplication of both wheat and barley chromosomes (Islam et
al., 1978, 1981). Islam et al. (1978, 1981) report that only one of 20 F_1 hybrids obtained possessed the 28 chromosomes
expected (Figure 5.7). The remaining 19 possessed from 21 to
36 chromosomes in different plants.

Meiotic

Abnormal chromosome counts are found in backcross progenies
of some intergeneric hybrids that possess the expected chromosome complement (Islam and Shepherd, 1980; Mujeeb-Kazi and
Rodriguez, 1981b). Chromosome banding techniques have shown
that such progeny can possess novel chromosome combinations
(Islam and Shepherd, 1980; Jewell, 1980). F_1 progeny from
T. aestivum x *Aegilops variabilis* possess 35 chromosomes; 21
wheat and 14 *Ae. variabilis* (Jewell, 1980). On backcrossing
to wheat unreduced gametes of the hybrid are expected to
yield progeny that possess 56 chromosomes (42 (21 II) wheat
and 14 (14 I) *Ae. variabilis*) (Figure 5.8). In addition to
progeny with the expected chromosome number, progeny with
chromosome counts of 39 to 63 were obtained. Similarly normal progeny of *H. vulgare* x *T. aestivum* backcrossed to wheat
are expected to possess 49 chromosomes (42 (21 II) wheat and

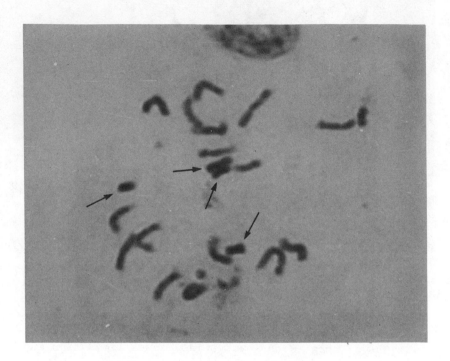

Figure 5.7. Non-classical maize x Tripsacum cytology, 16 maize and 4 Tripsacum (arrowed) chromosomes.

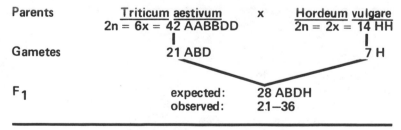

FIGURE 8. CHROMOSOME NUMBERS OF
T. AESTIVUM X H. VULGARE PROGENY

Figure 5.8. Chromosome numbers of T. aestivum x H. vulgare progeny.

7 (7 I) barley) (Figure 5.9). However, in addition, progeny
with 27 to 50 chromosomes have been obtained (Mujeeb-Kazi and
Rodriguez, 1981b); and in the reciprocal cross (*T. aestivum-*
H. vulgare x *T. aestivum*) progeny with 45 to 51 chromosomes
have been reported (Islam and Shepherd, 1980). N-banding has
shown that these backcross progeny from *T. aestivum* x *Ae.*
variabilis (Jewell, 1980) and from *T. aestivum* x *H. vulgare*
(Islam et al., 1981) possess abnormal chromosome complements
from both parents.

Such somatic and meiotic variations could extend the potential for new chromosome combinations immensely, although
the particular combinations that result may be unpredictable
with our present knowledge.

Karyotype

In addition to chromosome elimination and duplication,
karyotype variation has been observed in intergeneric hybrids
(Tsuchiya 1964; Lange and Jochemsen, 1976).

In some intergeneric hybrids there is a change in chromosome size of one genome, often in association with chromosome
elimination. This has been found after sexual hybridization
in *Nicotiana suaveolens* x *N. glutinosa* hybrids (Gupta and
Gupta, 1973) and after somatic cell fusion of *N. glauca* with
Glycine max (soybean) (Kao, 1977). In these examples a

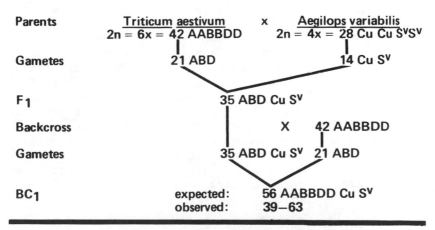

**FIGURE 9. CHROMOSOME NUMBERS OF
T. AESTIVUM X A. VARIABILIS PROGENY**

Figure 5.9. Chromosome numbers of T. aestivum x Ae.
variabilis progeny.

reduction in chromosome size was observed with successive
cell divisions. Mitotic disturbances and the presence of
chromosome bridges and fragments in hybrid cells suggested
that chromosome breakage was the cause.

Chromosome breakages and rearrangements of chromosome segments have been reported also in the nonclassical wheat x
barley hybrids (Shepherd and Islam, 1981). Such reassociation of genetic material can further extend the range of new
combinations of alien genetic material. Although, again the
result may be somewhat unpredictable.

Morphological changes in rye chromosomes have been observed when they are added to the wheat complement (Tarkowski
et al., 1972). Bhattacharyya et al. (1961) reported a change
in the relative arm lengths of one rye chromosome, and a disappearance of the secondary constriction in another. The
latter is a common phenomenon in intergeneric hybrids, and
has been reported in *Agropyron* (Heneen, 1962), *Avena*
(Sadasivaiah and Rajhathy, 1969), and *Hordeum* (Lange and
Jochemsen, 1976). However such karyotype modification has
not been associated with a change in genetic constitution,
nor with a loss in inheritance potential.

IV. POTENTIAL ALIEN GENOME UTILIZATION

Alien germplasm may be utilized to various degrees. The
success of *Triticale* shows that the combination of whole
genomes can give a new crop plant (Wolff, 1976). However
this success must be limited to plants buffered by their high
ploidy level and to the combination of closely compatible
genomes. Similarly the use of addition lines (O'Mara, 1940)
and substitution lines (Halloran, 1976) is limited to polyploids and to chromosomes with a high degree of compatibility.
In general the transfer of specific characters isolated on
small chromosome segments is required to avoid linked transfer of undesirable traits or other deleterious effects of the
alien chromatin (Knott, 1978). Alien genome combination is
the first step in this process.

Hybrid Fertility

For conventional transfer and utilization of alien genetic material, hybrids must retain at least part of the alien genome in a stable condition through functional gamete formation.

Most intergeneric hybrids are male sterile, but the production of functional female gametes is common. Often viable unreduced gametes are facultative gametophytic apomicts, in which case replicate F_1 plants are recovered in backcross progeny from intergeneric hybrids. This occurs with maize x *Tripsacum* hybrids, and with *Hordeum-Triticum* hybrids (Mujeeb-Kazi, 1981). However, these gametes also may function sexually, which after further crossing allows new chromosome combinations. Additional potential for novel chromosome combination is created by the production of functional gametes with alien chromosome combinations from mitotically variable hybrids, as found in some maize x *Tripsacum* hybrids (James, 1979), and in some wheat x barley hybrids (Islam et al., 1978, 1980), and from the meiotically variable hybrids mentioned earlier (Islam and Shepherd, 1980; Jewell, 1980) (Figure 5.10).

Some intergeneric hybrids are both male and female sterile, which severely restricts their utilization in plant improvement. In some cases it has been possible to induce fertility by producing the amphiploid of the hybrid with colchicine treatment (Chapman and Miller, 1978). However, many attempts at amphiploid production with intergeneric hybrids have been unsuccessful (Islam et al., 1978; Fedak, 1980; Mujeeb-Kazi and Rodriguez, 1981a).

Alien Genome Homoeology

The degree of homoeology between genomes of different species will influence greatly the potential use of alien germplasm for plant improvement (Kimber and Sallee, 1976). A lack of homoeology between genomes can prevent incorporation of germplasm from distantly related species (Galinat, 1977). However, this is true for certain genome combinations only, as introgression has been reported for an impressive range of alien germplasms. Examples are known from *Saccharum* into sorghum (de Wet et al., 1976), from *Tripsacum* into *Z. mays* (Harlan and de Wet, 1978; de Wet, 1979; Hooker, 1981), and from many different relatives into wheat (Knott, 1978).

When lack of homoeology is considered to be a serious problem, the desired transfer can be aided by bridge crosses (Knott, 1978) or irradiation techniques (Sears, 1956; Knott, 1956; Larter and Elliott, 1956). In wheat the transfer may be accomplished by utilizing the genetic control of chromosome

Jacqueline James Sprague

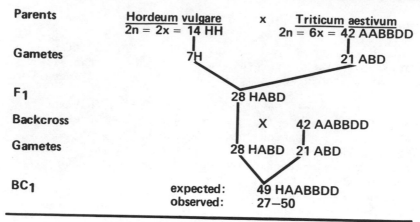

FIGURE 10. CHROMOSOME NUMBERS OF H. VULGARE X T. AESTIVUM PROGENY

Figure 5.10. Chromosome numbers of H. vulgare x T. aestivum progeny.

pairing (Riley, 1958; Sears and Okamoto, 1958; Riley et al., 1968; Sears, 1976).

In fact, lack of homoeology may be an advantage in some cases. Knott (1978) reported that disease resistance genes transferred to wheat from its distant relatives appear to be more effective and stable than those transferred from diploid and tetraploid species with homoeologous genomes.

Expression of Alien Germplasm

Ultimately the value of alien germplasm for crop improvement will depend upon obtaining a desired expression in agronomically suitable plant types. However germplasm can express itself differently in native and alien backgrounds. For example, although wheat-barley addition lines are morphologically different from the wheat parent, they do not exhibit any obvious morphological character from barley (Islam et al., 1978). Similarly the transfer of certain alien rust resistance genes into *T. aestivum* gives a reduction or loss of the resistance in the bread wheat background (Knott, 1978). On the other hand, there are examples of successful alien gene transfer, where the desired character is expressed in the new background. For example Hooker (1981) reported successful transfer of resistance to *Helminthosporium turcicum* from *T. floridanum* into maize. Also there are many examples of useful character transfer into wheat, not the least of which is the transference of *Agropyron* stem rust resistance which has been used successfully in Australia (Knott, 1978).

The potential use of alien germplasm depends on the specific germplasm combination involved. Sufficient genetic and/or environmental variation seems to exist in the barriers to crossability to permit the production of hybrids between many related species; and more successful intergeneric hybrids are produced each year as barriers to hybridization are overcome via new techniques.

Too few hybrids have been produced to date to permit any generalization on the potential of intergeneric crosses for plant improvement, but their analyses yield valuable information which may lead to a greater and more efficient utilization of such germplasm in the future.

SUMMARY

Alien genome combination is necessary to assess the potential contribution of alien germplasm in crop improvement. The production of hybrids by the conventional, sexual route is restricted by incompatibility barriers established during the course of evolution. Pollen germination and pollen tube growth must allow delivery of male gametes to the embryo sac for successful fertilization. Endosperm and embryo development must follow to produce a viable, fertile hybrid. In general, the more distantly related the parents, the more difficult it is to produce a hybrid between them. In many interspecific crosses, enough genetic variability exists to allow the production of a limited number of hybrids, and progress in methods to overcome incompatibility barriers allows the production of more hybrids from more widely divergent parents each year.

In addition to hybrid production, alien pollination sometimes stimulates the development of unreduced or reduced egg cells to give maternal plants and haploids respectively. Haploids are produced also by genome elimination after fertilization. Other chromosome variation in F_1 progeny gives rise to progeny with novel chromosome combinations. For conventional transfer of genetic material, hybrids must retain at least part of the alien genome through functional gamete formation. A greater control of alien chromosome behavior and gene expression would make possible selective retention and use of desirable germplasm from more genetic sources than is possible at present.

ACKNOWLEDGMENTS

The author wishes to thank CIMMYT staff for their valuable comments, and their help in the preparation of the paper.

REFERENCES

Baker, L. R., Chen, N. C., and Park, G. H.   (1975).   *HortScience 10*, 272.

Barclay, I. R.   (1975).   *Nature 256*, 410.

Bhattacharyya, N. K., Evans, L. E., and Jenkins, B. C.   (1961). *Nucleus* (Calcutta) *4*, 25.

Brink, R. A., and Cooper, D. C.   (1947).   *Bot. Rev. 13*, 423.

Caspari, E.   (1948).   *Adv. Genet. 2*, 1.

Chapman, V., and Miller, T. E.   (1978).   *Cer. Res. Comm. 6*, 351.

Davies, D. R.   (1958).   *Heredity 12*, 493.

Davies, D. R.   (1974).   *Heredity 32*, 267.

Davies, D. R., and Wall, E. T.   (1961).   *In* "Effects of Ionizing Radiation on Seeds," p. 83.   Int. Atom. Energy Agency, Vienna.

de Wet, J. M. J.   (1979).   *Proc. Conf. Broadening Genet. Base Crops*, p. 203.   Pudoc, Wageningen.

de Wet, J. M. J., Gupta, S. C., Harlan, J. R., and Grassl, C. O.   (1976).   *Crop Sci. 16*, 568.

Dhaliwal, S., and King, P. J.   (1978).   *Theor. Appl. Genet. 53*, 43.

Fedak, G.   (1977).   *Nature 266*, 529.

Fedak, G.   (1980).   *Can. J. Genet. Cytol. 22*, 117.

Galinat, W. C.   (1977).   *In* "Corn and Corn Improvement," p. 1. Academic Press, New York.

Gerrish, E. E.   (1967).   *Maize Genet. Coop. Newsl. 41*, 26.

Gupta, S. B.   (1969).   *Can. J. Genet. Cytol. 11*, 133.

Gupta, S. B., and Gupta, P.   (1973)   *Genetics 73*, 605.

Gustafsson, A.   (1946).   Lunds Univ. Arsskr. N. F. Ard. 2. 42, 1.

Halloran, G. M.   (1976).   *Euphytica 25*, 65.

Harlan, J. R., and de Wet, J. M. J.   (1978).   *Agron. Abstr.*, p. 53.

Heneen, W. K.   (1962).   *Hereditas 48*, 471.

Hooker, A. L.   (1981).   *Maize Genet. Coop. Newsl. 55*, 87.

Islam, A. K. M. R.   (1980).   *Chromosoma 76*, 365.

Islam, A. K. M. R., and Shepherd, K. W.   (1980).   *Chromosoma 76*, 363.

Islam, A. K. M. R., Shepherd, K. W., and Sparrow, D. H. B. (1975).   *Proc. 3rd Int. Barley Genet. Symp.* (Garcing, West Germany).   p. 260.

Islam, A. K. M. R., Shepherd, K. W., and Sparrow, D. H. B. (1978). *Proc. 5th Int. Wheat Genet. Symp.* (New Delhi). p. 365.

Islam, A. K. M. R., Shepherd, K. W., and Sparrow, D. H. B. (1981). *Heredity 46*, 161.

James, J. (1978). *Maize Genet. Coop. Newsl. 52*, 12.

James, J. (1979). *Euphytica 28*, 239.

Jewell, D. C. (1980). Ph.D. Thesis, Univ. of Adelaide.

Kao, K. N. (1975). *In* "Plant Tissue Culture Methods" (Gamborg and Welter, eds.), p. 22. Nat. Res. Council Canada.

Kao, K. N. (1977). *Molec. Genet. 150*, 225.

Kasha, K. J., and Kao, K. N. (1970). *Nature 225*, 874.

Kiesselback, T. A. (1949). *Nebr. Agr. Exp. Sta. Res. Bul. 161*.

Kimber, G., and Sallee, P. J. (1976). *Cer. Res. Comm. 4*, 33.

Knott, D. R. (1956). *Proc. 3rd Int. Wheat Rust Cong.* p. 72.

Knott, D. R. (1964). *Can. J. Plant Sci. 41*, 109.

Knott, D. R. (1978). *Proc. 5th Int. Wheat Genet. Symp.* (New Delhi). p. 354.

Kruse, A. (1973). *Hereditas 73*, 157.

Lange, W., and Jochemsen, G. (1976). *Genetica 46*, 217.

Lange, W., and Wojciechowska, B. (1976). *Euphytica 25*, 609.

Larter, E. N., and Elliott, F. C. (1956). *Can. J. Bot. 34*, 817.

Linskens, H. F. (1972). *Soviet Pl. Physiol. 20*, 156.

Livingston, G. K., and Stettler, R. F. (1973). *Rad. Bot. 13*, 65.

Mujeeb-Kazi, A. (1981). *J. Hered. 72* (in press).

Mujeeb-Kazi, A., and Rodriguez, R. (1980). *Cer. Res. Comm. 8*, 469.

Mujeeb-Kazi, A., and Rodriguez, R. (1981a). *J. Hered. 72* (in press).

Mujeeb-Kazi, A., and Rodriguez, R. (1981b). *J. Hered. 72* (in press).

Ohyama, K. (1975). *In* "Plant Tissue Culture Methods" (Gamborg and Wetter, eds.), p. 28. Nat. Res. Council Canada.

O'Mara, J. G. (1940). *Genetics 25*, 401.

Pandey, K. K. (1974). *Heredity 33*, 279.

Reger, B. J., and James, J. (1982). *Crop Sci. 22* (in press).

Riley, R. (1958). *Proc. X Int. Cong. Genet. 2*, 234.

Riley, R., Chapman, V., and Johnson, R. (1968). *Genet. Res. 12*, 199.

Sadasivaiah, R. S., and Rajhathy, T. (1969). *Can. J. Genet. Cytol. 10*, 655.

Sears, E. R. (1944). *Genetics 29*, 113.

Sears, E. R. (1956). *In* "Genetics in plant breeding." *Brookhaven Symp. Biol. 9*, 1.

Sears, E. R.   (1976).   *Annu. Rev. Genet. 10,* 31.

Sears E. R., and Okamoto, M.   (1958).   *Proc. X Int. Cong. Genet. 2,* 258.

Shepherd, K. W., and Islam, A. K. M. R.   (1981).   *In* "Wheat Science--Today and Tomorrow" (Evans and Peacock, eds.), p. 107.   Camb. Univ. Press.

Siddiqui, K. A., and Jones, J. K.   (1969).   *Euphytica 18,* 71.

Snape, J. W., Bennett, M. D., and Simpson, E.   (1980).   *Z. Pflanzenzüchtg. 85,* 200.

Stebbins, G. L.   (1950).   "Variation and Evolution in Plants." Columbia Univ. Press, New York.

Stebbins, G. L.   (1958).   *Adv. Genet. 9,* 147.

Subrahmanyam, N. C., and Kasha, K. J.   (1973).   *Chromosoma 42,* 111.

Taira, T., and Larter, E. N.   (1977).   *Can. J. Bot. 55,* 2330.

Tarkowski, C., and Stefanowska, G.   (1972).   *Genet. Pol. 13,* 83.

Tozu, T.   (1966).   *Seiken Ziho 18,* 33.

Tsuchiya, T.   (1964).   *Proc. 1st Int. Barley Genet. Symp.* p. 116.

Tuzuka, N.   (1952).   *Bul. Res. Inst. Food Science,* Kyoto University.   p. 8.

Vijendra Das, L. D.   (1970).   *J. Hered. 61,* 288.

Wolff, A.   (1976).   Cimmyt Today No. 5.

Zenkteller, M., and Straub, J.   (1979).   *Z. Pflanzenzüchtg. 82,* 36.

Zeven, A. C., and Van Heemert, C.   (1970).   *Euphytica 19,* 175.

CHAPTER 6

USE OF EMBRYO CULTURE IN INTERSPECIFIC HYBRIDIZATION

Elizabeth G. Williams
Isabelle M. Verry
Warren M. Williams

Grasslands Division, D.S.I.R.
Palmerston North
NEW ZEALAND

I. INTRODUCTION

Interspecific hybridization offers plant breeders a
method for increasing the range of variation available within
a pasture or crop species, and for introducing desirable
genetic factors such as disease resistance into cultivated
species. Natural barriers to interspecific hybridization may
occur before or after fertilization. For example, pollen may
fail to penetrate a foreign pistil, or two distantly related
genomes may be unable to produce a viable embryo when combined
as a zygote. However, the use of interspecific hybridization
between closely related species is often limited by postfertilization failure of endosperm development and consequent
abortion of seeds carrying potentially viable hybrid embryos.
Rescue of these embryos by sterile *in vitro* culture is now a
well established technique (see for example, Williams and de
Lautour, 1980; Rupert and Evans, 1980a,b; Phillips *et al.*,
1980; Reed and Collins, 1980).

II. PREVIOUS EMBRYO CULTURE RESULTS

In previous work with the pasture legume genera
Trifolium, Lotus and *Ornithopus* we have used transplanted
normal nurse endosperm from a parental or other leguminous

plant to supplement the artificial nutrient medium and
provide a more natural immediate environment for young hybrid
embryos in culture. The methodology has been described in
detail by Williams and de Lautour (1980). Hybrids raised by
the nurse endosperm technique include *Trifolium ambiguum* M.
Bieb. (4x) x *T. repens* L. (Williams, 1978; Williams and Verry,
1981); *T. ambiguum* (4x) x *T. hybridum* L. (2x) (Williams,
1980; *T. repens* x *T. uniflorum* L. (Pandey, Grant, Petterson
and Williams - in preparation); *Lotus pedunculatus* Cav. x
L. corniculatus L. (de Lautour, unpubl.; Verry, unpubl.;
Williams and Williams, 1981); *L. pedunculatus* (4x) x *L.
tenuis* Waldst. et Kit (4x) (de Lautour, Jones and Ross, 1978;
L. tenuis (4x) x *L. corniculatus* (de Lautour, Jones and Ross,
1978; *Ornithopus sativus* brot. x *O. compressus* L. (Williams
de Lautour, Plummer and Williams - in preparation); *O.
pinnatus* (Mill.) Druce x *O. sativus* (Williams and de Lautour,
1981); *Triticum durum* Desf. x *Secale cereale* L. (Williams and
Driscoll - unpubl. internal report to DSIR, New Zealand, and
Waite Agricultural Research Institute, South Australia, 1981).

A.   *T. ambiguum* (4x) x *T. repens*

    T. repens (white clover) is a stoloniferous perennial
which establishes rapidly and produces well in pasture. However, its productivity is lowered by several virus diseases,
and its shallow fibrous root system contributes to its
susceptibility to drought and root-chewing insects. *T.
ambiguum* (Caucasian clover, Kura clover) is a hardy, cold-
tolerant, more deeply rooting, rhizomatous perennial,
resistant to several viruses which attack white clover
(Barnett and Gibson, 1975) but slow to establish in pasture.
Hybridization might ideally combine the shoot vigour of *T.
repens* with the virus resistance and below-ground vigour of
T. ambiguum.
    From several hundred hand pollinated flowers four F_1
hybrid plants have been raised. Three are highly sterile and
one partially fertile (and self-compatible) with about 23%
normal pollen. One sterile hybrid shows unusually high
flowering intensity, while another flowers only at daylengths
greater than 16 hours. Flowering responses of the third
sterile hybrid and the partially fertile hybrid are more
normal.
    The partially fertile hybrid and several of its F_2
progeny have performed well as spaced plants in grazed field
trials. A few F_3 and backcrosses to *T. repens* have also been
produced. Most individuals in the hybrid population show

pollen fertility below 10% and have the stoloniferous habit of
T. repens. However, several appear to be rhizomatous, and a
number are strongly rooting.

B.   *T. ambiguum* (4x) x *T. hybridum* (2x)

Hybrids between these species have been obtained
previously by Keim (1953) and Evans (1962) using embryo
culture. *T. hybridum* (alsike clover) is a tall, single-crown
plant which is agriculturally useful in the South Island high
country of New Zealand, but very susceptible to virus diseases
(Ashby, 1976). Hybridization is aimed at combining the tall
shoot of *T. hybridum* with the virus resistance and underground
spreading habit of *T. ambiguum*.
    So far, from several hundred hand pollinations we have
produced two highly sterile triploid F_1 hybrids carrying two
genomes from *T. ambiguum* and one from *T. hybridum*. The other
is performing well as spaced plants in grazed field trials,
and shows a stronger resemblance to *T. ambiguum* with vigorous
rooting and a lower and more spreading crown than *T. hybridum*.
    Future success of this cross will depend on obtaining
fertile hybrids, either by crossing the parental species at
the 2x or 4x level, or by colchicine doubling of the chromosome number of sterile triploids. Both approaches are
currently being attempted.

C.   *T. repens* x *T. uniflorum*

Hybrids between these species have been obtained
previously by Pandey (1957) and Gibson *et al.* (1971) without
embryo culture. *T. uniflorum* is an agriculturally unimportant,
single-crown species which lacks the productivity and spreading stoloniferous habit of *T. repens* but is deeply tap-rooted.
The main aim of the cross is to confer a deeper rooting habit
on *T. repens*, and thereby increase resistance to drought and
root-chewing insects.
    The hybrid population, which includes F_1 and F_2 plants
and backcrosses to *T. repens*, generally shows the prostrate,
small-leaved, tap-rooted habit of *T. uniflorum*. Some
individuals show stolon development with deep tap-rooting
from the nodes. Since fertility is variable but generally
low, initial selection is being carried out for improved
fertility, but there are prospects for selection of productive
plants with a deeply rooting habit.

D.   *L. pedunculatus* x *L. corniculatus*

L. corniculatus is a well established agricultural
species which produces well at high soil fertility under
intensive farming. *L. pedunculatus*, however, although slow
to recover from defoliation (Brock and Charlton, 1977), can
be used on soils of low fertility and pH (Levy, 1970), is
resistant to chewing insects (Farrell and Sweney, 1974), and
contains foliar tannins which help to prevent bloat in
grazing ruminants (Jones and Lyttleton, 1971).
Several F_1 hybrids were produced in 1962 by embryo
culture using tetraploid *L. pedunculatus* derived by colchicine
treatment of the diploid (de Lautour, unpubl.). A selected F_3
population has been evaluated as the breeder's line
'Grasslands 4712', and although variable in type, predominantly resembles *L. corniculatus* in habit. Plants contain foliar
tannins (Ross and Jones, 1974), but lack resistance to chewing
insects (Farrell and Sweney, 1974) and are inferior in yield
and persistence to recently evaluated populations of *L.
corniculatus* (Charlton *et al.*, 1978). The genetic basis of
this hybrid population is now being extended by production of
new F_1 hybrids between improved cultivars of tetraploid *L.
pedunculatus* ('Grasslands Maku') and *L. corniculatus*
('Franco'). Embryos are being successfully cultured on half-
strength Eriksson's (1965) medium without nurse endosperm
(Verry, unpubl.).

E.   *L. pedunculatus* (4x) x *L. tenuis* (4x) and *L. tenuis* (4x)
      x *L. corniculatus*

These hybrids were produced to study the inheritance of
condensed tannins (Jones *et al.*, 1976; de Lautour *et al.*,
1978) and are not currently being used for further breeding.

F.   *O. sativus* x *O. compressus*

O. sativus (pink serradella) is productive and of
potential agronomic significance as a winter annual pasture
legume on sandy soils in New Zealand (Williams *et al.*, 1975).
However, it is soft-seeded and therefore unable to build
quantities of buried dormant seed important for consistent
year to year performance. It is also upright in habit and

susceptible to grazing damage. *O. compressus* lacks the vigour
of *O. sativus* and its seeds are unhullable, but it is hard-
seeded and prostrate in habit. A single F_1 hybrid was
obtained by partial ovule culture with transplanted *Ulex
europaeus* endosperm. Although this hybrid was morphologically
abnormal, with less than 10% pollen fertility, it has given
rise to an F_6 breeder's line which shows high fertility and
variation for the characters under selection. A number of
families in the population combine all the desired features of
the two parent species.

G.   *O. pinnatus* x *O. sativus*

This cross was made to combine the prostrate growth habit
and foliar tannins of *O. pinnatus* with the higher productivity
of *O. sativus*. Four highly sterile diploid F_1 hybrids were
obtained by embryo culture. One was chromosome-doubled with
colchicine and attained partial fertility (less than 10%
fertile pollen) but no viable seeds were produced owing to
endosperm failure. Embryo culture with nurse endosperm was
repeatedly required to obtain F_2 to F_5 generations because
of continuing endosperm failure. A few seeds showing almost
normal endosperm and embryo development were obtained on
several F_4 plants suggesting that selection for normal seed
set might be possible in future generations. Apart from seed
inviability, most plants of the F_2 to F_5 hybrid populations
were vegetatively vigorous and combined the desired characteristics of the parent species.

H.   *T. durum* x *S. cereale*

The lines of *Triticale*, derived from crosses of wheat x
rye, which are currently in use or under evaluation in
Australia, have been derived from CIMMYT Mexican materials
selected under irrigation and high nitrogen conditions.
Although these lines suit the Australian seasons, they are
poorly adapted to infertile soils and moisture limitation.
Crossing of tetraploid durum wheat (*T. durum*) with South
Australian Commercial rye (*S. cereale*) which has been
selected under dry, marginal conditions, should provide
triticales better suited to Australia's low rainfall areas.
In the cross between tetraploid durum wheats and rye,
endosperm failure prevents the formation of viable seed *in*

situ, and embryo culture is required to raise hybrid plants.
Bajaj (1980) found that growth of these hybrid embryos was
enhanced by culturing them in contact with a layer of macerated, immature wheat endosperm spread on the surface of an
artificial nutrient medium, a technique similar to our nurse
endosperm transplant method.

There have been occasional reports that vegetative grafting of related but sexually incompatible species enhances their
ability to produce hybrid offspring (see for example, the work
of Nirk (1959) in *Lycopersicon*).   Evans (1960) suggested that
such results were probably not caused by grafting *per se*, but
reflected a correlation between vegetative and sexual compatibility, such that success of grafts acted as a selective
screen for genotypes which were also sexually compatible.
However, Hall (1954), using a technique developed by
Pissarev and Vinogradova (1954), found that grafting of wheat
embryos on to rye endosperms at germination enhanced the
ability of the resulting wheat plants to hybridize with rye
at maturity.   Grafting wheat embryos from a line incompatible
with rye on to endosperm of a more compatible wheat also
enhanced crossability with rye (Hall, 1956).

We have made a pilot-scale test of grafting at germination
and embryo culture with nurse endosperm for the durum wheat x
rye cross, using South Australian adapted parent materials.
Wheat plants grafted at germination to rye endosperms were
male sterile and required no emasculation before pollination
with rye pollen.   From only 29 flowers pollinated, seven
hybrid embryos were obtained.   These were cultured on intact,
immature hexaploid wheat endosperms on nutrient medium EC6
(B. and N. Williams and de Lautour, 1980, plus Ca pantothenate
0.1 mg/l, glutamine 40 mg/l, adenine sulfate 0.001 mg/l,
kinetin 0.01 mg/l; NAA 0.1 mg/l, Agar 0.7%; pH 5.7).   Five
embryos showed some growth of shoot and/or root, and one
hybrid was transferred successfully to soil.

The use of intact or macerated nurse endosperm was tested
in a separate pilot-scale experiment using normal wheat and
rye embryos ca. 1 mm in length.   Embryos grown without nurse
endosperm tended to germinate prematurely to give small,
weak seedlings.   Contact with either intact or macerated
nurse endosperm appeared to prolong embryonic differentiation
and prevent premature germination.   Macerated endosperm
caused initial callus-like growth of embryos although later
differentiation was normal and more vigorous than on intact
endosperm.

Further tests will be required to confirm these observa-
tions for both grafting at germination and embryo cultue.
However, large-scale production of primary triticales using
South Australian adapted materials appears feasible.

III. CURRENT EMBRYO CULTURE PROGRAMS

Although the nurse endosperm technique has been applied
successfully in the above crosses, the double dissection
procedure requiring one normal endosperm for each hybrid
embryo is time-consuming. In addition, growth on the culture
media currently in use has been much slower than seed
maturation *in vivo*, and many hybrid embryos have died in
culture after the initiation of growth. Our current work is
therefore centred around better definition of the nutritional
requirements of young legume embryos in culture so that a
rate of growth and development can be obtained to parallel
that occurring *in vivo*, without requiring nurse endosperm. We
have also simplified some technical manipulations, to increase
the volume of material handled and allow routine processing
by technical staff.

Media modifications being systematically tested include
several major salt formulations and a number of concentrations
of sucrose, osmotically active additives, B vitamins, casein
hydrolysate, yeast extract, IAA, NAA and cytokinins. Four
size classes of normal *T. repens* embryos between 0.15 and 0.50
mm are being used without nurse endosperm in these tests.
Developmental stages range from early "hearts" with the first
tinge of green to small green "torpedos". Our most successful media have had low major salt concentrations (e.g. half-
strength Eriksson's (1965) or half-strength PC-LZ from
(Phillips and Collins, 1979)), 4% sucrose and 0.25 to 3.0 g/l
yeast extract. Although some abnormalities in early developmental patterns may occur, young *T. repens* embryos grow at
approximately the normal *in vivo* rate on such media. Growth
through to seedlings can be obtained, but transfer to a
medium with lower sucrose (2%) and yeast (0.125 g/l) levels
after 7-10 days may facilitate later transfer to a potting
mixture. Data are also being collected which show that the
cultivars used in hybridization, and also the genotypes of
individual pairs of parent plants, are important in determining the extent of embryo growth *in vivo* and *in vitro* and
eventual success of the cross. These results will be
reported in detail elsewhere.

Although modifications are still assessed we are using
these media to grow 0.15 to 0.7 mm hybrid embryos from the
cross *T. semipilosum, Lotus, Medicago sativa* and *Capsella
bursa-pastoris*, and will be used in the coming flowering
season to attempt to raise hybrids from the crosses *T.
ambiguum* (2x) x *T. hybridum* (2x), *T. ambiguum* (4x) x *T.
hybridum* (4x), and to obtain further hybrids of *T. ambiguum*
(4x) x *T. repens*.

IV.   CONCLUSION

The triticales are now agriculturally established
products of interspecific hybridization. However, hybridization of forage legumes has not yet produced commercial cultivars, and although interspecific hybridization has been extensively used in pasture grasses (Carnahan and Hill, 1961), few
commercial hybrid cultivars have been released. Problems of
infertility or instability of hybrid populations (Stebbins,
1958) frequently limit the usefulness of this approach. Apart
from *Lolium perenne* L. x *L. multiflorum* Lam. hybrids (Corkill,
1945), which are fully fertile, the commercially released
hybrid grasses suffer from fertility problems. For example,
Phalaris tuberosa L. x *P. arundinacea* L. hybrids are sterile
pentaploids and F_1 seed is produced each year by exploitation
of self-incompatibility to achieve interspecific fertilization
(McWilliam, 1974).

Among the legume hybrids discussed in this paper, two,
Ornithopus sativus x *O. compressus* and *Lotus pendunculatus* x
L. corniculatus, appear to be fully fertile and suffer from
no fertility barrier to their potential success and eventual
release as hybrid cultivars. Two others, *T. ambiguum* x *T.
repens* and *T. repens* x *T. uniflorum*, show reduced fertility
but give some indication of prospects for useful genetic
exchange between the parent species and for improvement in
fertility by selection. Triploid sterility in the cross
T. ambiguum (4x) x *T. hybridum* (2x) and endosperm failure
causing seed inviability in tetraploid hybrids of *O. pinnatus*
x *O. sativus* are more serious barriers to gene exchange.

We believe that improvement of the embryo culture
technique will, for suitable crosses, allow the production of
extensive hybrid populations comprising F_1 and **advanced**
generation hybrids and backcrosses to parental species. Such
variable gene pools might be maintained to allow continuing
genetic exchange with periodic selection for desired
combinations of parental characters. Prospects for generating
new variation and transferring genetic material between
closely related species appear to be excellent even where
initial hybrid fertility is low.

ACKNOWLEDGMENTS

We are grateful to our colleagues Dr D.W.R. White and
G. de Lautour for advice and assistance, to Professor C.J.
Driscoll, Waite Agricultural Research Institute, South

Australia, for wheat and rye materials and advice, and to Professor R.B. Knox, School of Botany, University of Melbourne, Australia, for facilities to carry out part of this work during the tenure of a University of Melbourne Research Fellowship award to E.G. Williams.

REFERENCES

Ashby, J.W. (1976). *N.Z. J. Agric. Res. 19*, 373.
Bajaj, Y.P.S. (1980). *Cereal Res. Comm. 8*, 359.
Barnett, O.W., and Gibson, P.B. (1975). *Crop Sci. 15*, 32.
Brock, J.L., and Charlton, J.F.L. (1977). *Proc. N.Z. Grassld Assoc. 39*, 121.
Carnahan, H.L., and Hill, Helen D. (1961). *Bot. Rev. 27*, 1.
Charlton, J.F.L., Wilson, E.R.L., and Ross, M.D. 1978). *N.Z. J. Exo. Agric. 6*, 201.
Corkhill, L. (1945). *N.Z. J. Agric. 71*, 465.
de Lautour, G., Jones, W.T., and Ross, M.D. (1978). *N.Z. J. Bot. 16*, 61.
Eriksson, T. (1965). *Physiol. Plant 18*, 976.
Evans, A.M. (1960). *Rep. Welsh Plant Breeding Stn 959*, 81.
Evans, A.M. (1962). *Euphytica 11*, 164.
Farrell, J.A.K., and Sweney, W.J. (1974). *N.Z. J. Agric. Res. 17*, 69.
Gibson, P.B., Chen, C.-C., Gillingham, J.T., and Barnett, O.W. (1971). *Crop Sci. 11*, 895.
Hall, O.L. (1954). *Hereditas 40*, 453.
Hall, O.L. (1956). *Hereditas 42*, 261.
Jones, W.T., Broadhurst, R.B., and Lyttleton, J.W. (1976). *Phytochemistry 15*, 1407.
Jones, W.T., and Lyttleton, J.W. (1971). *N.Z. J. Agric. Res. 14*, 101.
Keim, Wayne, F. (1953). *Agron. J. 45*, 601.
Levy, E.B. (1970). Government Printer, Wellington, New Zealand.
McWilliam, J.R. (1974). *Proc. XII Int. Grass. Congr. 3(2)*, 901.
Nirk, H. (1959). *Nature 184*, 1819.
Pandey, K.K. (1957). *J. Hered. 48*, 278.
Phillips, G.C., and Collins, G.B. (1979). *Crop Sci. 19*, 59.
Phillips, G.C., Collins, G.B., and Taylor, N.L. (1980). *Abs. Amer. Soc. Agron.* 65.
Pissarev, W.E., and Vinogradova, N.M. (1944). *Comptes Rendus (Doklady) Acad. Sci. USSR 45 No.3.*
Reed, Sandra M., and Collins, G.B. (1980). *Abs. Amer. Soc. Agron.,* 66.

Ross, M.D., and Jones, W.T. (1974). *N.Z. J. Agric. Res. 17*, 191.

Rupert, E.A., and Evans, P.T. (1980a). *Abs. Amer. Soc. Agron.*, 66.

Rupert, E.A., and Evans, P.T. (1980b). *Clovers and Special Purpose Legumes Research 13*, 65.

Stebbins, G.L. (1958). *Adv. Genetics 9*, 147.

Williams, E. (1978). *N.Z. J. Bot.*, *16*, 499.

Williams, E.G. (1980). *N.Z. J. Bot.*, *18*, 215.

Williams, E.G., and de Lautour, G. (1980). *Bot. Gaz. 141*, 252.

Williams, E.G., and de Lautour, G. (1981). *N.Z. J. Bot. 19*, 23.

Williams, E.G., and Verry, I.M. (1981). *N.Z. J. Bot. 19*, 1.

Williams, W.M., de Lautour, G., and Stiefel, W. (1975). *N.Z. J. Exp. Agric. 3*, 339.

Williams, W.M., and Williams, E.G. (1981). *Proc. XIV Int. Grasslands Congress, Kentucky*. (In press).

CHAPTER 7

HAPLOIDS IN PLANT IMPROVEMENT

Chih-ching Chu

Institute of Botany
Academia Sinica
Peking

I.   INTRODUCTION

Since Blakeslee *et al.* (1922) discovered the first
haploid plant in *Datura*, haploid sporophytes, spontaneous or
derived from experimental   treatment *in vivo*, have been
reported in many species of angiosperms (Kimber and Riley,
1963).   In 1964, the induction of haploid embryos from pollen
of *Datura innoxia* by anther culture was successfully achieved
(Guha and Maheshwari, 1964; 1966).   This work greatly
stimulated the interest in induction of androgenetic haploids.
Attention had also been paid to other methods of haploid
production, for example, by chromosome elimination and ovary
culture.   In the last decade the progress on haploid
production has led to extensive studies for the utilization of
haploidy in plant breeding and some new cultivars of tobacco,
rice and wheat have been selected from doubled haploids
(Anonymous, 1974b; Yin *et al.*, 1976; Anonymous, 1976b).   The
recent advances on haploid research have been reviewed by
several authors (Sunderland, 1977; Nitzsche and Wenzel, 1977;
Collins, 1977; Sink, 1977; Keller and Stringham, 1978; Vasil,
1980).   In this paper the emphasis will be placed upon the
production of haploids and their application to plant
breeding.   The references published in Chinese will be
discussed in some detail here.

II.   PRODUCTION OF HAPLOIDS

Generally, there are two pathways to produce haploid

plants: *in vivo* and *in vitro*. The development of *in vivo* haploids proceeds within the ovary, thus they may also be called *in ovulo* haploids (Vasil, 1980). It is generally assumed that *in vivo* haploid sporophytes are derived from some component of the embryo sac. However, in a few instances they can arise from male gametes in female cytoplasm (Kimber and Riley, 1963). *In vivo* haploid sporophytes can be induced from either male gametophytes by anther culture or female gametophytes by ovary culture.

A.   In vivo *Production of Haploids*

In vivo haploid embryos can be produced from the ovary after treatment by one of the following procedures: delayed pollination, pollination with irradiated pollen, alien pollination and temperature shocks. In most cases the induction frequency of haploids is rather low. But in potato the frequency of parthenogenetic haploids via 4X-2X crosses can be greatly increased since the discovery of the superior pollinators in the diploid *Solanum phureja* (Hougas and Gabert, 1964; Montelongo-Escobedo and Rowe, 1969). Such a superior pollinator was also found in *Zea mays* (Kermicle, 1969).

A more efficient procedure of haploid production in barley is based on chromosome elimination. Symko (1969) and Kasha and Kao (1970) reported that a large number of haploid plants could be obtained from the hybrid seed between *Hordeum vulgare* and *H. bulbosum* by using embryo culture technique. Further investigation indicated that the haploid plants were produced from the hybrid embryos in which the chromosomes of *H. bulbosum* had been eliminated in the early developing stage (Subrahmanyan and Kasha, 1973). The chromosome elimination method was also successfully adapted from barley to wheat (Barclay, 1975).

In cotton, semigamy as an efficient method is used for producing haploid plants. Barrow and Chaudhari (1976) found that when the F_1 hybrids of *Gossypium hirsutum* X *G. barbadense* as a male parent were crossed with a virescent semigametic strain, chimeras with virescent (female) and green (male) foliage were produced among the progenies. When the haploid green sectors were treated with colchicine, they set homodiploid seeds.

B.   Anther Culture

As anther culture is a chief method to produce haploids, it will receive more attention in this discussion.

1. Species in which Haploids have been Obtained. Up to now haploid pollen plants have been obtained in at least 111 species or hybrids. Most of them belong to *Solanaceae, Gramineae* and *Cruciferae*. It is worth while to mention the successful induction of pollen plants of several woody genera, e.g. *Populus, Hevea, Citrus, Poncirus, Vitis* and *Aesculus*.

Table 1. Species in which pollen Plants have been obtained

Species	*References*
Aegilops caudata X A. umbellulata	*Kimata and Sakamoto (1972)*
Aesculus hippocastanum	*Radojevic (1978)*
Anemone canadensis	*Johansson and Eriksson (1977)*
Anemone coronaria	*Sunderland and Dunwell (1977)*
Anemone hepatica	*Georgiev and Chavdarov (1974)*
Anemone multifida	*Johansson and Eriksson (1977)*
Anemone rupicula	*Johansson and Eriksson (1977)*
Anemone virginiana	*Johansson and Eriksson (1977)*
Aradidopsis thaliana	*Gresshoff and Doy (1972b)*
Asparagus officinalis	*Pelletier et al. (1972)*
Atropa belladonna	*Zenkteler (1971a,b)*
Brassica campestris	*Keller et al. (1975)*
Brassica chinensis	*Zhong et al. (1978)*
Brassica napus	*Thomas and Wenzel (1975b)*
Brassica oleracea	*Kameya and Hinata (1970)*
Brassica oleracea X B. alboglabra	*Kameya and Hinata (1970)*
Brassica pekinesis	*Teng and Kuo (1978)*
Capsicum annuum	*Y.Y. Wang et al. (1973)*
Capsicum frutescens	*Novak (1974)*
Catharanthus roseus	*Abour-Mandour (1979)*
Citrus microcarpa	*Z.G. Chen et al. (1980)*
Coix lacryma	*C.C. Wang et al. (1980)*
Datura innoxia	*Guha and Maheshware (1964,,1966)*
Datura metel	*Narayanaswamy and Chandy (1971)*
Datura meteloides	*Kohlenbach and Geier (1972) Nitsch (1972)*

Species	References
Datura muricata	*Nitsch (1972)*
Datura stramonium	*Guha and Maheshware (1967)*
Datura wrightii	*Kohlenbach and Geier (1972)*
Digitalis purpurea	*Corduan and Spix (1975)*
Festuca pratensis	*Zenkteler and Misiura (1974)*
Glycine max	*Yin et al. (1980)*
Hevea brasiliensis	*C.H. Chen et al. (1978,1979)*
Hordeum vulgare	*Clapman (1973)*
Hyoscyamus albus	*Raghavan (1975)*
Hyoscyamus niger	*Corduan (1975), Raghavan (1975)*
Hyoscyamus pusillus	*Raghavan (1975)*
Ipomoea batatas	*Sehgal (1978), Tsay and Tseng (1979)*
Lilium longiflorum	*Sharp et al. (1972)*
Lolium multiflorum	*Clapham (1971)*
Lolium multiflorum X Festuca arundinacea	*Nitzsche (1970)*
Lolium perenne	*Clapham (1971)*
Lycium barbarum	*Gu (1981)*
Lycium halimifolium	*Zenkteler (1972)*
Lycopersicum esculentum	*Gresshoff and Doy (1972b)*
Lycopersicum pimpinellifolium	*Debergh and Nitsch (1973)*
Nicotiana affinis	*Nitsch and Nitsch (1969)*
Nicotiana alata	*Nitsch (1969)*
Nicotiana attenuata	*Collins and Sunderland (1974)*
Nicotiana clevelandii	*Vyskot and Novak (1974)*
Nicotiana glauca X N. langsdorffii	*Smith (1974)*
Nicotiana glutinosa	*Nitsch (1969)*
Nicotiana knightiana	*Collins and Sunderland (1974)*
Nicotiana otophora	*Collins et al. (1972)*
Nicotiana langsdorffii	*Durr and Fleck (1980)*
Nicotiana paniculata	*Nakamura et al. (1974b)*
Nicotiana raimondii	*Collins and Sunderland (1974)*
Nicotiana rustica	*Nitsch and Nitsch (1969)*
Nicotiana sanderae	*Vyscot and Novak (1974a)*
Nicotiana suaveolens	*Smith (1974)*
Nicotiana sylvestris	*Bourgin and Nitsch (1967)*
Nicotiana tabacum	*Bourgin and Nitsch (1967)*
Nicotiana tabacum X N. sylvestris	*Takahashi (1973)*
Oryza perennis	*Wakasa and Watanabe (1979)*
Oryza sativa	*Niizeki and Oono (1968)*
Oryza sativa X O. perennis	*Woo et al. (1978)*

Species	References
Pelargonium hortorum	*Abo El-Nil and Hildebrant (1971)*
Petunia axillaris	*Doreswamy and Chack (1973), Engvild (1973)*
Petunia hybrida	*Iyer and Raina (1972), Binding (1972)*
Petunia hybrida X P.	*Raquin and Pilet (1972)*
Pharbitis nil	*Sangwan and Norreel (1975a)*
Physalis minima	*George and Rao (1979)*
Poncirus trifoliata	*Hidaka et al. (1979)*
Populus alba X P. simonii	*Chang (1978a,b)*
Populus berolinensis	*Lu et al. (1978a,b)*
Populus berolinensis X P. pyramidalis	*Lu et al. (1978a,b)*
Populus canadensis X P. koreana	*Anonymous (1977a)*
Populus euphratica	*Zhu et al. (1980)*
Populus euramericana	*Zhu et al. (1980)*
Populus harbinensis X P. pyramidalis	*Anonymous (1977a)*
Populus nigra	*C.C. Wang et al. (1975)*
Populus pekinensis	*Lu et al. (1978a,b)*
Populus pseudo-simonii	*Lu et al. (1978a,b)*
Populus psuedo-simonii X P. pyramidalis	*Lu et al. (1978a,b)*
Populus simonii	*Lu et al. (1978a,b)*
Populus simonii X P. nigra	*Anonymous (1975a)*
Populus simonii X P. pyramidalis	*Lu et al. (1978a,b)*
Populus ussuriensis	*Anonymous (1975a)*
Rehmannia glutinosa	*Anonymous (1979)*
Saccharum officinarum	*C.H. Chen et al. (1980)*
Saintpaulia ionantha	*Hughes et al. (1975)*
Scopolia carniolica	*Wernicke and Kohlenbach (1975)*
Scopolia lurida	*Wernicke and Kohlenbach (1975)*
Scopolia physaloides	*Wernicke and Kohlenback (1975)*
Secale cereale	*Thomas and Wenzel (1975a), Thomas et al. (1975)*
Secale montanum	*Zenkteler and Misiura (1974)*
Setaria italica	*Ban et al. (1971)*
Solanum dulcamara	*Zenkteler (1973)*
Solanum melongena	*Iyer and Raina (1972), Raina and Iyer (1973)*
Solanum nigrum	*Harn (1972)*
Solanum phureja	*Matsubayashi and Kuranuki (1975)*
Solanum surattense	*Sinha et al. (1979)*

Species	References
Solanum tuberosum	Dunwell and Sunderland (1973)
Solanum verrucosum	Irikura and Sakaguchi (1972)
Sorghum vulgare	Anonymous (1978a)
Triticale (6X)	Bernard et al. (1976)
Triticale (8X)	Sun et al. (1973), Y.Y. Wang et al. (1973)
Triticum aestivum	Ouyang et al. (1973), Chu et al. (1973), C.C. Wang et al. (1973), Picard and Buyser (1973)
Triticum durum	Chu et al. (1979)
Triticum aestivum X Agropyron glaucum	C.C. Wang et al. (1975b)
Vitis vinifera	Zhou and Li (1981)
Zea mays	Anonymous (1975c)

Since 1971, Chinese botanists have made considerable effort to induce the pollen plants of many economically important species. They have reported regeneration of pollen plants from 32 species of cereals, vegetables and trees for the first time. More recently, the induction of pollen plants has been achieved in sugar cane, *Citrus, Vitis* and *Coix*. Since these references have been available only in Chinese or in the form of Chinese papers with short English abstracts, the main results are listed in Table 2.

2. *The Factors Influencing Induction Frequency of Pollen Plants.* For practical breeding purposes a high frequency of haploid production is necessary. To increase the induction frequency various factors influencing the response of anthers in culture have been critically studied.

a. *Genotype of donor plants.* The ability of pollen to form haploid sporophytes is obviously affected by the genotypic differences among the donor plants because the frequency of pollen plants varies with genus, species, subspecies and cultivar. Most species of *Nicotiana* produce pollen plants readily, but in *N. langsdorffii* only a few pollen embryos could be induced on a special medium during certain growth season (Durr and Fleck, 1980). In rice, the subspecies *japonica* always shows higher yield of pollen plants than the subspecies *indica* (Anonymous, 1975b; Guha-Mukhergel, 1973). Cultivar to cultivar variation in anther response to culture

Table 2. Methods of anther culture in some species*

Species	Induction of pollen calli	Plant regeneration	References
Saccharum officinarum	N + 2,4-D 2 mg/l + kinetin 2 mg/l + LH 500 mg/l, sucrose 20%	MS + 6-BA 2 mg/l	C.H. Chen et al. (1980)
Citrus microcarpa (cv. Sijiju)	N + 2,4-D 2 mg/l + 6-BA 2 mg/l (or + 2,4-D 0.5 mg/l + kinetin 2 mg/l), sucrose 8%	MS + 6-BA 2 mg/l + IAA 0.1 mg/l + LH 500 mg/l, sucrose 2%	Z.G. Chen et al. (1980)
Vitis vinifera (cv ShenLi)	Modified B + 6-BA 2 mg/l + 2,4-D 0.5 mg/l, sucrose 3%	Modified B + 6-BA 4 mg/l + NAA 0.2 mg/l (or MS + 6-BA 0.5 mg/l), sucrose 3%	Zhou and Li (1981)
Coix lacryma	N + 2,4-D 2 mg/l, sucrose 3%	N basic medium or N + kinetin 0.5 mg/l + NAA 1 mg/l, sucrose 3%	C.C. Wang et al. (1980)

*The anthers at uninucleate stage of microspores were used for culture in every species.

can be detected in nearly all species tested. In the extreme
instances, as seen in *Sorghum vulgare, Vitis vinifera, Citrus
microcarpa* and *Zea mays*, the production of pollen plants are
confined to one or a few cultivars (Anonymous 1978a; Zhou and
Li, 1981; Z.H. Chen et al., 1980; Anonymous, 1975c). Picard
and Buyser (1977) reported that a higher frequency of pollen
plants occurred in the microspore-derived lines than in the
control. A further increase in the frequency was obtained
from the donor plants by crossing certain microspore-derived
lines. Chinese investigators also noticed that the hybrids
of which one parent showed high yield of pollen plants often
produced more pollen plants (Ouyang *et al.*, 1973; Kuo *et al.*,
1978). This indicates that genes or gene recombination may
affect androgenesis (Jacobson and Sopory, 1978). In fact,
Zamir *et al.* (1980) demonstrated that a single gene mutation
in tomato could influence the induction of pollen callus and
subsequent regeneration of plants. It has not been known
whether the genes directly control the formation of pollen
embryos. As the lower yield of pollen embryos in a certain
cultivar of tobacco can be increased by changing the light
condition of anther culture (Phillips and Collins, 1977), it
is likely that each genotype requires special culture
conditions to induce pollen embryogenesis.

　　　b. Physiological state of donor plants. The physiological
state of donor plants can considerably influence the frequency
of pollen plants. In tobacco, the higher yield of pollen
plants is obtained when anthers are collected from the
earliest developing flower buds. Meanwhile the highest level
of embryogenesis is observed when donor plants are grown
under short days and high light intensity (Dunwell, 1976),
According to Sunderland (1978), the anthers derived from the
nitrogen-starved plants of tobacco are far more productive
than contrast plants. In rape seed, the donor plants grown at
low temperatures gave the greatest embryo yield, whereas
higher frequencies of barley or wheat callus are produced from
the plants grown at higher temperatures in late summer
(Foroughi-Wehr *et al.*, 1976; Wang and Chen, 1980).

　　　c. Developmental stage of pollen. Although tetrads and
mature pollen can develop into pollen plants (Nakata and
Tanaka, 1968; Abo El-Nil and Hildbrandt, 1971,1973;
Nitzsche, 1970; Kameya and Hinata, 1970), the uninucleate
microspores are more suitable for induction of pollen plants.
Usually, the microspores just before or at mitosis show the

best response to anther culture in tobacco (Nitsch, 1969;
Sunderland, 1970) and rice (C.C. Wang *et al.*, 1974; Anonymous,
1974a). In wheat and maize, however, the younger uninucleate
microspores may be superior for induction of pollen callus
(Ouyang *et al.*, 1973; Kuo *et al.*, 1978; Cheng *et al.*, 1978).

   d. *Culture media*. Nitsch H medium (Nitsch and Nitsch,
1969) and MS medium have usually been used with success for
anther culture of dicotyledons. In cereal crops, earlier
studies on anther culture were carried out on MS, Miller or LS
medium (Niizeki and Oono, 1968; Chu *et al.*, 1973; Clapham,
1971). Later, Clapham (1973) and Chu *et al.* (1975) found that
the higher concentration of ammonium ion could inhibit formation of pollen calli in barley and rice. Based on this
result, Chu *et al.* (1975) developed N6 medium which has
proved to be more efficient than other synthetic media for
anther culture of rice and other cereals (Chu, 1978; Genovesi
and Majil, 1979; Sun *et al.*, 1980; Nitzsche and Wenzel, 1977;
Kuo *et al.*, 1978; Anonymous, 1977c). In addition, a simple
medium consisting of 20% aqueous potato extract has been
developed to culture the anthers of tobacco, rice and wheat
(Anonymous, 1976a, 1977b). The potato medium gave far better
results than MS medium. However, its use leads to experimental variability depending on potato cultivars and their
physiological state. To minimise this variability, a modified
potato medium which was supplemented with major elements, iron
salt and thiamine was devised by Chuang *et al.* (1978) and
proved to be more effcient for wheat.
   The hormones in the medium are critical factors for
embryo or callus formation. Kinetin or other cytokinins are
necessary for induction of pollen embryos in many species of
Solanaceae except tobacco (Guha and Maheshwari, 1967; Sopory
and Maheshwari, 1976; Sopory *et al.*, 1978). Auxin, particularly 2,4-D, can greatly promote the formation of pollen
callus in cereals (Chu *et al.*, 1978) and make the pollen
embryos develop into calli in some species of Solanaceae
(Narayanaswamy and Chandy, 1971; Sunderland and Wicks,
1971; Geier and Kohlenbach, 1973; Rashid and Street, 1973;
Raghavan, 1978). For regeneration of plants from pollen
calli, a cytokinin and lower concentration of auxin are often
needed. But in rice and wheat, plant differentiation on
hormone-free medium is as good as on the hormone-containing
ones (Guha *et al.*, 1970; Chu *et al.*, 1976).
   Sucrose has been considered to be the most effective
carbohydrate source which can not be substituted by other
disaccharides (Keller *et al.*, 1975; Sangwan-Norreel, 1977).
Glucose can be used in anther culture in some cases (Wenzel

et al., 1977), but fructose is far less effective (Keller, 1978). The concentrations of sucrose also play an important role in induction of pollen plants. In tobacco and rice, 2-5% sucrose is suitable for both callus formation and plant regeneration (Nitsch and Nitsch, 1969; Kuo *et al.*, 1973; Chu, 1978). Other species, such as potato, rapeseed, wheat, Triticale and barley, require higher levels of sucrose, generally ranging from 6-12% (Sopory *et al.*, 1978; Keller *et al.*, 1975; Chu *et al.*, 1978; Sun *et al.*, 1980; Clapham, 1973). In maize, the best response of anthers to culture occurs on the medium with 12-15% of sucrose (Anonymous, 1975, 1966c; Miao *et al.*, 1978; Ku *et al.*, 1978). The sucrose concentration for anther culture of sugarcane is as high as 20% (C.H. Chen *et al.*, 1980).

To improve efficiency, activated charcoal has been used in anther culture of tobacco, rapeseed, potato, maize, Triticale and rye (Anagnostakis, 1973; Wenzel *et al.*, 1977; Sopory *et al.*, 1978; Miao *et al.*, 1978; Sun *et al.*, 1980). Kohlenbach and Wernicke (1978) pointed out that the effect of activated charcoal on increasing the yield of pollen plants is likely due to the removal of inhibitors from the agar. This explanation is confirmed by the fact that on liquid medium, which does not contain the inhibitors from agar, the frequency of pollen embryos or pollen calli can be greatly increased (Wernicke and Kohlenback, 1976; Wilson, 1977; C. Hu *et al.*, 1978; Sunderland, 1978). In addition activated charcoal can absorb 5-hydroxymethylfutaral, a product of sucrose dehydration during autoclaving, which is assumed to be an inhibitor of growth in anther culture of tobacco (Weatherhead *et al.*, 1978).

Certain organic supplements are beneficial to anther culture, including casein and lactalbumin hydrolysates (C.C. Wang *et al.*, 1974; Pan *et al.*, 1978; C.H. Chen *et al.*, 1980), coconut milk (Sun *et al.*, 1973), nucleic acid hydrolysate (Chu *et al.*, 1973), yeast extract (C.C. Wang *et al.*, 1974), ascorbic acid and glutathione (Wenzel *et al.*, 1977).

     e. Physical factors of culture. The temperature of culture has an influence on the frequency of pollen callus in *Datura* and wheat. The anthers cultured at 30°C always produce more pollen calli than those at 25°C or lower temperature (Sopory *et al.*, 1976; Ho *et al.*, 1978). When rapeseed anthers are cultured at elevated temperatures prior to transfer to 25°C, the yields of embryos is sharply increased (Keller and Stringham, 1978). As to the light factor, contradictory results have been reported depending on the

species tested. In *Datura*, the inhibitory effects of continuous light on pollen embryogenesis is marked (Sopory and
Maheshwari, 1976). The light is also unfavorable for production of pollen embryos in Triticale. In contrast to *Datura*
and Triticale, continuous light greatly enhances the yield of
pollen embryos in some tobacco cultivars (Phillips and Collins,
1977). Few studies have dealt with the effect of light
quality except the inhibition of red light on pollen embryogenesis of *Datura* (Sopory and Maheshwari, 1976).

   f. Cold pretreatment. Nitsch and Norrel (1973) discovered that cold treatment of flower buds of *Datura* prior to
culture greatly enhanced the yield of haploid plants. This
has been widely confirmed in other species. Table 3 shows the
proper temperature and duration of cold pretreatment in
different species.

Table 3. Procedures of cold treatment of flower buds

Species	Temperature	Duration	References
Secale cereale	6°C	3-15 days	Nitzche and Wenzel (1977)
	1-3°C	7-14 days	Sun et al.(1978)
Datura innoxia	3°C	48 h	Nitsch and Norreel (1973)
Datura metel	0-1°C	48 h	Gupta and Babbar (1980)
Hordeum vulgare	3°C	48 hr	Bouharmont(1977)
Nicotiana tabacum	7-9°C	7-14 days	Sunderland(1978)
Oryza sativa	10°C	48 h	Wang et al.(1974)
	10-13°C	7-14 days	Genovesi and Magill (1979)
Petunia hybrida	6°C	48 h	Malhotra and Maheshware(1977)
Triticum aestivum	3-5°C	48 h	Pan et al. (1975)
Triticale	3-5°C	72 h	Sun et al. (1980).

   Nitsche and Norreel (1973) suggested that chilling of
anthers prior to culture disrupted the normal asymmetric pollen
mitosis, resulting in an increased frequency of pollen with two
equal sized cells. Such pollen grains were considered to be
the source of pollen embryos. However, Sunderland (1978) has
been able to obtain an increase of pollen with two equal

nuclei by cold pretreatment. He explained that the effect of
cold treatment was due to delayed pollen deterioration. This
controversial problem should be solved by further experiments.

　　　g. *Pollen dimorphism*. A dimophic population of pollen
has been found in several species. In these species, a few
small anomalous pollen, termed S-grain, can be observed
besides the large normal pollen population (Horner and Street,
1978). The S-grain is characterized by the small size and a
poor affinity for cytoplasmic stains and usually undergoes
extra division, similar to those described in *in vitro*
cultured anthers. It has been suggested the pollen embryos
arise only from the S-grains and the pollen embryogenesis in
cultured anthers is predetermined before the excision of the
flower buds (Dale, 1975; Horner and Street, 1978; Sunderland,
1978). This point of view is supported by the result that the
frequency of embryogenetic pollen grains is not increased by
in vitro induction in tobacco (Horner and Mott, 1979).
Contrary results reported by Wilson *et al*. (1978) and Dunwell
(1978) indicate that the number of embryogenetic grains can be
increased over and above the number of S-grains. It seems
that the hypothesis of predetermination of pollen embryogenesis
should be examined by more experimental results.

　　　3. *Ploidy of Pollen Plants*. The majority of pollen plants
in tobacco are haploid (Bourgin and Nitsch, 1967; Sunderland,
1971). However, nonhaploid, including diploid, polyploid,
mixoploid and aneuploid, have been found in nearly all species
tested (McComb, 1978). It is well known that the proportion
of nonhaploids varies with the age of pollen at the time of
excision. In *Datura* and *Nicotiana*, the diploid and polyploid
pollen increase when older anthers are cultured (Narayanaswamy
and Chandy, 1971; Engvild, 1973,1974; Engvild *et al*., 1972).
On the other hand, there may be a relationship between the
proportion of haploids and the constituents of hormones in the
media since in other systems the auxins have marked influence
on ploidy of culture cells (Binding, 1974).
　　　In a few cases the diploid plants derived from cultured
anthers may develop from parental somatic tissues (Niizeki
and Grant, 1971) or unreduced microspores (Wenzel *et al*.,
1976). However, most of diploid pollen plants are really
microspore-originated. The best evidence for this come from
the genetic analysis of diploid plants derived from the
anthers of F_1 hybrids which differ for many characters but

their progenies are very uniform (Niizeki and Oono, 1971;
Ouyang *et al.*, 1973; Yin *et al.*, 1976;   Hsu *et al.*, 1975).
The spontaneous diploids and polyploids among pollen plants
generally have been attributed to endomitosis and nuclear
fusion in cultured pollen (Raquin and Pilet, 1972;   Engvild,
1973,1974; Chu *et al.*, 1978; Hu *et al.*, 1978). The
aneuploids and mixoploids are assumed to be caused by
irregular mitosis of pollen callus (Mix and Foroughi-Wehr,
1978; H. Hu *et al.*, 1978).
     For breeding purposes, the haploids should be diploidized.
Chromosome doubling can be achieved by treatment with colchicine or phenylmercuric-*p*-toluenesulfonanilide, a cheaper
germicide (Yin *et al.*, 1976). Recently, DMSO (dimethyl
sulfoxide), as a adjuvant of colchicine, has been successfully
used for increasing the efficiency of chromosome doubling
(Kaul and Zutshi, 1971; Subrahmanyam and Kasha, 1975; Bai *et
al.*, 1979). Meanwhile, the homozygous diploids can also be
produced from somatic cultures of haploid pollen plants
(Nitsch *et al.*, 1969a; Kasperbauer and Collins, 1971; Kohhar
et al., 1970; Niizeki *et al.*, 1978; Weatherhead and Henshaw,
1979).

     4. Albino Pollen Plants. Albinos are a common phenomenon
in pollen plants of Gramineae. The frequencies of albinos
range from 5% to 90%, depending on cultivar and culture
temperature. The temperature of culture can obviously
influence the frequency of albinos. When there is a rise in
temperature of anther culture, the frequency of albinos is
increased (C.C. Wang *et al.*, 1977; Bernard, 1980). Except
culture temperature, the other cultural factors *in vitro*
appears to have no effect on albino frequency.
     The cause of albinism is an unsolved problem. Since the
albino plants do contain plastids (Clapham, 1973; Sun *et al.*,
1974), they cannot come from plastid-free generative cells.
The lack of 23S and 16S rRNA as well as fraction I protein in
albino plants shows that albinism may be caused by a variation
of the plasmid DNA (Sun *et al.*, 1978b). Other possible
causes of albinism have been proposed: (1) albino plants
originate from pollen with chromosome aberration (Chu *et al.*,
1978; (2) reorganization of plastids fails following the
regression of proplastids during meiotic development (Nilson-
Tillgren and von Wettstein-Knowles, 1970); (3) physiological
factors operating *in vitro* (Bernard, 1980).

C.   *Isolated Pollen Culture*

The production of haploids by using isolated pollen
culture has two advantages: (1) obtaining large quantity of
haploid plants from fewer anthers; (2) avoiding the inter-
ference of somatic callus developed from anther wall.
The earlier attempts to culture pollen grains of *Brassica*
(Kameya and Hinata, 1970), *Lotus* (Niiseki and Grant, 1971) and
Petunia (Binding, 1972) resulted in only a few cell divisions
and the same is true of microspore culture of rye (Wenzel *et
al.*, 1975). However, successful reports have gradually
increased in this field. At present isolated pollen grains
have been grown to plants in *Datura innoxia* (Nitsch and
Norrell, 1973), *Nicotiana tabacum* (Nitsch, 1974), *Solanum
tuberosum* (Sopory, 1977), *Hyoscyamus* (Wernicke and Kolenback,
1977), *Petunia hybrida* (Sangwan and Norrell, 1975), *Atropa*
(Sunderland and Roberts, 1977), *Solanum melongena* (Gu, 1979)
and *Oryza sativa* (Y. Chen *et al.*, 1980). In most cases only
those pollen grains which are derived from precultured
anthers can develop into embryos. This shows that the anther
wall may play a critical role during the early stage of
culture. Also, the addition of an aqueous extracts of
cultured embryogenic anthers promotes the embryogenesis
of isolated pollen (Nitsch and Norreel, 1973; Debergh and
Nitsch, 1973). Further experiments suggest that the active
compounds in the extracts may be cytokinins and serine (Nitsch,
1974). On the other, hand embryogenesis and plant formation of
isolated pollen grains derived from the unprecultured anthers
have been reported in *Petunia* and *Solanum tuberosa* (Sangwan
and Norreel, 1966; Sopory, 1977).

D.   *Ovary Culture*

To investigate test tube fertilization and fruit development, much attention have been paid to ovary culture
(Rangaswamy, 1977). More attention is now being given to
female gametophyte culture in gymnosperms and unfertilized
ovary culture in angiosperms for inducing haploid embryogenesis. The haploid embryos or plants produced from *in
vitro* cultured female gametophytes have been reported in
Zamia floridana (La Rue, 1948), *Zamia integrifolia* (Nostog,
1965), and *Ephedra foliata* (Konar and Singh, 1979). In
angiosperms, more recent reports indicate that haploid
plants can be induced from cultured unfertilized ovaries of

Hordeum vulgare (San Noreum, 1976,1979;   C.C. Wang and Kuang,
1981), *Nicotiana tabacum, Triticum aestivum* (Zhu and Wu, 1979),
Oryza sativa (Zhou and Yang, 1980,1981). The histological
investigations revealed that the haploids originated from
embryo sac elements. In *Hordeum*, the egg or antipodal cells
can develop into embryos, whereas the synergids gave only
proliferation of callus (San Noreum, 1979). The haploid
embryos of rice also originate from the egg apparatus, but it
has not been determined whether they come from egg cells or
synergids. At present there is no evidence for the origin of
haploid embryos from antipodals or polar nuclei, although they
may undergo limited cell divisions (Zhou and Yang, 1981).

The haploid plants from ovaries may be obtained either
directly on a medium with lower level of auxin, or regenerated
from calli after transferring them to the differentiating
medium (Table 4). The ovaries at younger embryo sac stage,
when the pollen are at late uninucleate or binucleate stage,
usually give the best response to induction of haploid plants
(Zhou and Yang, 1980,1981). It is also worthy to note that
the ovaries with attached receptacles and stamen always
produce more haploid plants than those without them.

III. APPLICATION OF HAPLOIDY TO BREEDING

A.   *Production and Utilization of Homozygous Lines*

Soon after the discovery of the first haploid plant,
Blakeslee and Belling (1924) indicated that the homozygous
diploid could be obtained from haploid by means of chromosome
doubling. Through haploids homozygosity can be achieved
even in self-incompatible, dioecious and vegetatively propagated species. Once established, homozygous lines have
potential for plant improvement in a number of ways.

First,   the haploid technique can be employed in a
cross breeding program. After doubling the haploids derived
from microspores of hybrids, the homozygous lines can be
immediately produced. In this case, generations of inbreeding,
which are necessary in conventional breeding procedure, can
be omitted and the breeding cycle for new cultivars can be
greatly shortened.

Another advantage of the haploid technique lies in higher
selection efficiency compared with that of diploid breeding.
Theoretically, if the parents of hybrid have n pairs of
independent recombining alleles, to select a particular

Table 4. Induction of haploids from cultured ovaries

Species	Induction of Plants		References
	Induction of calli	Regeneration of plants	
Oryza sativa	N6 (solid or liquid) + MCPA 0.125 mg/L, sucrose 3%	Decreasing MCPA* to 0.033 mg/L	Zhou and Yang (1981)
Triticum aestivum	N6 + 2,4-D 2 mg/L, sucrose 8%	Avoiding 2,4-D, +IAA 0.2 mg/L + kinetin 1 mg/L, sucrose 2%	Zhu and Wu (1979)
Hordeum vulgare	Miller + Fujii (1970) vitamins + Fe-EDTA 10 mg/L, 2,4-D 2 mg/L, sucrose 10%	Avoiding 2,4-D, sucrose 2%	San Noreum (1976, 1979)
	N6 + 2,4-D 0.5 mg/L + NAA 1 mg/L + kinetin 1 mg/L, sucrose 3%		Wang and Kuang (1981)
Nicotiana tabacum	Nitsch H + IAA 0.5 mg/L + kinetin 2 mg/L, sucrose 2%		Zhu and Wu (1979)

*MCPA: 2,methyl-chlorophenoxyacetic acid.

genotype from the F_2 population, the selection efficiency should be $(1/2)^{2n}$ in haploid breeding and $(1/2)^n$ in diploid breeding.   This means that the selection efficiency in haploid breeding is 2^n times higher than that in diploid breeding. Owing to this reason, some new cultivars have been efficiently selected from smaller populations of pollen plants of tobacco (Anonymous, 1974b;   Nakamura et al., 1974b), rice (Yin et al., 1976;   Anonymous, 1976b) and wheat (Tsun, 1978).   In Brassica napus and eggplant, some doubled haploid lines have also been tested under field conditions (Keller and Stringham, 1978; Anonymous, 1978b).   Although the advantages of haploid breeding have been proved by practice, there still exists some divergence of opinion on the vitality of the doubled haploid lines.   In self-pollinated species, such as rice and wheat, the doubled haploid lines have no degeneration in vitality (Hsu et al., 1975; Hu et al., 1979).   But in tobacco which is partly hetero-gametic, the degeneration in vitality can sometimes be observed (Burk et al., 1973;   Oka et al., 1976) and sometimes not (Anonymous, 1974c).

Secondly, the doubled haploid lines are available for the utilization of heterosis in those heterogametic species.   It is well known that the homozygosity in doubled haploid lines is higher than that in inbred lines (Chase, 1969).   Such high purity of doubled haploids make it possible to produce the hybrids with maximum heterosis.   According to Wu et al. (1980) the hybrids between doubled haploid lines and inbred lines are often superior to those between inbred lines in their yield, growth vigor and uniformity.   In addition, the haploid technique can also be useful in the production of hybrid seeds in self-incompatible species, such as turnip, rape and Brassica oleracea (Keller et al., 1975;   Keller and Stringham, 1978). For poplars and other heterogametic trees, the haploid technique is especially valuable.   Because of a generation cycle of many years, the production of inbred lines by repeated selfings is nearly impossible in these species. Using the haploid technique, the production of homozygous lines and utilization of heterosis might be achieved (Lu et al., 1978a,b).

B.   Application of Haploidy to Distant Hybridization

The haploid technique, especially anther culture, is very useful in a breeding program utilising distant hybridization. With respect to this problem, the following theoretical possibilities can be advanced.

   *1. Obtaining Stable Fertile Lines from Hybrids between
Subspecies*. It is known that the F_1 hybrid between the rice
subspecies, japonica and indica, is partially sterile and that
the progeny segregate vigorously. However, the segregation of
the hybrid may be effectively controlled by anther culture.
Hsu *et al.* (1978) reported that the pollen-derived spontaneous
diploid from cultured anthers of *japonica-indica* hybrids is
homozygous and more than 50% of them are wholly fertile. A
new cultivar has been selected from the fertile lines and
released into commercial production.

   2. Transferring of Genes from Species to Species. If an
interspecific hybrid, in which the chromosomes can partially
pair in meiosis, is used for anther culture, some homozygous
lines with heterozygous translocation chromosomes could be
produced. In such cases, some genes responsible for good
quality, yield and resistance may be transferred from one
species to another.

   3. Production of Chromosome Additions and Substitutions.
O'Mara's method has been used for production of alien chromosome additions. It includes a cross between an amphidiploid
and one of its parents following selfing and selection
(O'Mara, 1940). When the anther culture technique is
introduced into O'Mara's procedure the production of addition
lines could be simplified. The *Triticum-Secale* hybrid is
used as an example to explain the experimental scheme (Fig.
1).
   On the other hand, anther culture provides a new
approach to produce chromosome substitutions from distant
hybrids. When *Triticum aestivum* is crossed with *Secale
cereale*, the dihaploid F_1 hybrid contains 21 univalent wheat
chromosomes and 7 univalent rye chromosomes. Owing to the
random distribution of the univalents during meiosis the
microspores of the hybrid have various number of chromosomes
which can be formulated nIW+nIR(0-28). Thus various chromosome substitutions could be obtained from microspores of the
hybrid by means of anther culture (Fig. 2).
   Nitzche and Wenzel (1977) tried to produce all substitutions from the haploid hybrid of *Festuca pratensis* X *Lolium
multiflorum* using the anther culture method. They obtained
4 albino pollen plants by culturing almost 18000 anthers.
Chu (1978) cultured about 10000 anthers of dihaploid hybrids
of *Triticum aestivum* X *Secale cereale* and obtained 2 calli,
which could not, however, differentiate into plants. These
preliminary results show that the production of chromosome

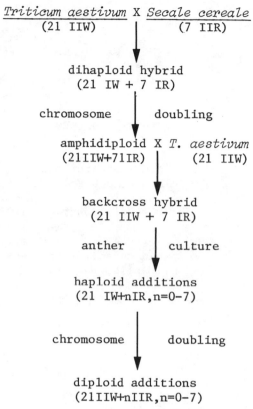

*Fig. 1. Chromosome addition breeding scheme of
          Triticum-Secale*

substitutions through anther culture though difficult is
possible. In contrast, the chromosome additions are easy
to induce by anther culture (Chu, 1978).

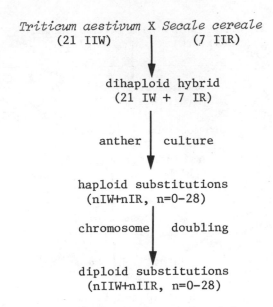

Fig. 2. Substitution Breeding Scheme of Triticum-Secale.

C.   *Mutagenesis of Haploids*

Haploidy is especially useful for the study of mutagenesis. Because there is only one set of chromosomes in the haploid and no masking of dominant allelic recessive relationships, once a gene mutation takes place, it is immediately expressed.   Induced mutagenesis has been carried out in pollen-derived haploid plants of tobacco (Nitsch, 1969; Nitsch *et al.*, 1969).   However, treating and screening a large number of whole plants is resource consuming. A more ideal experimental system for muta-genesis is haploid cell cultures, including callus, protoplasts, microspores and isolated cells. Using these experimental systems, microbiological techniques can be introduced into plant breeding. The successful results along this line have grown steadily during the past ten years.   Induced mutants have been selected from irradiated microspores in cultured anthers (Nitsch, 1972; Xuan *et al.*, 1978).   Somatic cell mutants which are resistant to antibiotic drugs (Binding *et al.*, 1970; Maliga *et al.*, 1973a,b), amino acid analogues (Chaleff and Carlson,

1974; Widholm, 1977) and herbicides (Chaleff and Parsons, 1978) have also been reported. In addition, Carlson (1970) isolated six autotrophic mutants by treating suspensions of tobacco haploid cells with EMS and then BUdR. Recent reviews of this field have been presented by Widholm (1977), Scowcroft (1977) and Thomas *et al.* (1979). Although a number of encouraging progresses have been made, the selection of haploid mutants is confined to a few species in which the haploid cultured cells can retain their haploidy and be induced to regenerate plants. Unfortunately, for most important crop plants the culture of haploid cells is unsuccessful. Wu and Zhong (1980) recently found some pollen callus lines in maize which could retain their haploidy and regenerate plants after a series of subcultures. They have obtained more than 20,000 haploid plantlets from the callus lines during about two years. These callus lines may be very useful for maize mutant breeding.

IV. CONCLUSIONS

The haploid technique has great potential for plant improvement. Some successful results indicate that the conventional breeding could be replaced by haploid breeding in self-fertilising crop plants, if a great number of doubled haploid lines can be produced. Instead of inbred lines, doubled haploid lines can be widely used for production of heterosis, if the haploids can be induced from various genotypes in a species. It is obvious that the successful utilization of haploids in plant breeding mainly depends upon the induction frequency of haploids. Thus it is important to study the mechanism of androgenesis and parthenogenesis *in vitro* for the production of haploids at high frequency. Another attractive subject is mutagenesis of haploid cells. It can be expected that more and more creative and fruitful contributions will grow up in this active research field.

REFERENCES

Abo El-Nil, M.M., and Hildebrant, A.C. (1971). *Amer. J. Bot.* *58*, 475.
Abo El-Nil, M.M., and Hildebrandt, A.C. (1973). *Can. J. Bot.* *54*, 2107.

Abour–Mandour, A.A., Fischer, S., and Czygan, F.C. (1979).
    Z. Pflanzenphysiol. 91, 83.
Anagnostakis, S.L. (1974). *Planta 115*, 281.
Anonymous (1974a). *Sci. Sin. 17*, 207.
Anonymous (1974b). *Acta Bot. Sin. 16*, 301.
Anonymous (1974c). *Acta Genet. Sin. 1*, 301.
Anonymous (1975a). *Sci. Sin. 18*, 769.
Anonymous (1975b). *Acta Genet. Sin. 2*, 81.
Anonymous (1975c). *Acta Genet. Sin. 2*, 138.
Anonymous (1976a). *Acta Genet. Sin. 3*, 25.
Anonymous (1976b). *Acta Genet. Sin. 3*, 19.
Anonymous (1977a). *Acta Genet. Sin. 4*, 49.
Anonymous (1977b). *Acta Genet. Sin. 4*, 302.
Anonymous (1977c). *Acta Genet. Sin. 4*, 361.
Anonymous (1978a). *Acta Genet. Sin. 5*, 337.
Anonymous (1978b). *Acta Bot. Sin. 20*, 154.
Anonymous (1979). *Bot. Gaz. (Peking), 4*, 8.
Bai, S.X., Liu, C.Z., and Zhang, C.Z. (1979). *Acta Genet.
    Sin. 6*, 230.
Ban, Y., Kokubu, T., and Miyaji, Y. (1971). *Bull. Fac. Agr.,
    Kagoshima Univ. 21*, 77.
Barclay, I.R. (1975). *Nature (Lond.), 256*, 410.
Barrow, J., and Chaudhari, H. (1976). *Crop Sci. 16*, 441.
Bernard, S. (1980). *Z. Pflanzenzuchtg. 85*, 308.
Bernard, S., Picard, E., and de Buyser, J. (1976). *C.R. Acad.
    Sci. 283 D*, 235.
Binding, H. (1972a). *Nature (Lond.) 237*, 283.
Binding, H. (1972b). *Z. Pflanzenzucht. 67*, 33.
Binding, H. (1974). *In* "Haploids in Higher Plants: Advances
    and Potential: (K.J. Kasha, ed.), Univ. of Guelph Press.
Binding, H., Binding, K., and Straub, J. (1970). *Naturwissen-
    schaften 57*, 138.
Blakeslee, A.F., Belling, J., Farnham, M.E., and Bergner, A.D.
    (1922). *Science 55*, 646.
Blakeslee, A.F., and Belling, J. (1924). *J. Hered. 15*, 195.
Bouharmont, J. (1977). *Caryologia 30*, 351.
Bourgin, J.P., and Nitsch, J.P. (1967). *Ann. Physiol. Veg. 9*,
    377.
Burk, L.G., Gwynn, G.R., and Chaplin, L.F. (1973). *J. Hered.
    63*, 355.
Carlson, P.S. (1970). *Science 168*, 487.
Chaleff, R.S., and Carlson, P.S. (1974). *Ann. Rev. Plant
    Physiol. 8*, 267.
Chaleff, R.S., and Parsons, M.A. (1978). *Proc. Nat. Acad.
    Sci. USA 75*, 5104.
Chang, L.K. (1978a). *In* "Proceedings of Symposium on Anther
    Culture: (H. Hu, ed.), p. 193. Science Press, Peking.

Chang, L.K. (1978b). *In* "Proceedings of Symposium on Plant Tissue Culture" p.241. Science Press, Peking.

Chase, S.S. (1969). *Bot. Rev. 35*, 117.

Chen, C.H., Chen, T.F., Chien, C.F., Wang, C.H., Chang, S.J., Hsu, H.E., Ou, H.H., Ho, Y.T., and Lu, T.M. (1978). *Acta Genet. Sin. 5*, 99.

Chen, C.H., Chien, C.F., Chen, M., Shuo, C.K., Wang, C.H., and Chen, F.C. (1980). *Hereditas (Beijing) 2*, 40.

Chen, Y., Wang, R.F., Tian, W.Z., Zuo, Q.X., Zheng, S.W., Lu, D.Y., and Zhang, G.H. (1980). *Acta Genet. Sin. 7*, 46.

Chen, Z.G., Wang, M.Q., and Liao, H.H. (1980). *Acta Genet. Sin. 7*, 189.

Cheng, W.C. Kuo, L.C, Kuan, Y.C., Huang, C.H., An, H.P., and Gu, M.K. (1978). *Acta Genet. Sin. 5*, 274.

Chu, C.C., Wang, C.C., Sun, C.S., Chien, N.F., Yin, K.C., and Hsu, C. (1973). *Acta Bot. Sin. 15*, 1.

Chu, C.C., Wang, C.C., Sun, C.S., Hsu, C., Yin, K.C. Chu, C.Y., and Bi, F.Y. (1975). *Sci. Sin. 18*, 659.

Chu, C.C., Wang, C.C., and Sun, C.S. (1976). *Acta Bot. Sin. 18*, 239.

Chu, C.C., Sun, C.S., and Wang, C.C. (1978). *Acta Bot. Sin. 20*, 6.

Chu, C.C. (1978). *In* "Proceedings of Symposium on Plant Tissue Culture" p.43. Science Press, Peking.

Chu, C.C., Wang, C.C., and Sun, C.S. (1979). *Acta Bot. Sin. 21*, 295.

Chuang, C.C., Ouyang, T.W., Chia, H., Chou, S.M., and Ching, C.K. (1978). *In* "Proceedings of Symposium on Plant Tissue Culture" p.51. Science Press, Peking.

Clapham, D. (1971). *Z. Pflanzenzuchtg. 65*, 285.

Clapham, D. (1973). *Z. Pflanzenzuchtg. 69*, 142,

Collins, G.B. (1977). *Crop. Sci. 17*, 583.

Collins, G.B., Legg, P.D., and Kasperbauer, M.J. (1972). *J. Hered. 63*, 113.

Collins, G.B., and Sunderland, N. (1974). *J. Exp. Bot. 25*, 1013.

Corduan, G. (1975). *Planta 127*, 27.

Corduan, G., and Spix, C. (1975). *In* "Genetic Manipulations with Plant Material" (L. Ledoux, ed.) p.576. Plenum Press.

Dale, P.J. (1975). *Planta 127*, 213.

Debergh, P., and Nitsch, C. (1973). *C.R. Acad. Sci. 276D*, 1281.

Doreswamy, R., and Chacko, E.K. (1973). *Hort. Res. 13*, 41.

Dunwell, J.M., and Sunderland, N. (1973). *Euphytica 22*, 317.

Dunwell, J.M. (1976). *Env. Exp. Bot. 16*, 109.

Dunwell, J.M. (1978). *In* "Frontiers of Plant Tissue Culture, 1978," (T.A. Thorpe, ed.), p.103. Univ. of Calgary Press, Calgary, Alberta.

Durr, A., and Fleck, J. (1980). *Plant Sci. Lett. 18*, 75.

Engvild, K.C. (1973). *Hereditas 74*, 144.

Engvild, K.C. (1974). *Hereditas 76*, 320.

Engvild, K.C., Linde-Laursen, I.B., and Lundquist, A. (1972). *Hereditas 72*, 331.

Foroughi-Wehr, B., Mix, G., Gaul, H., and Wilson, H.M. (1976). *Z. Pflanzenzuchtg. 77*, 198.

Fujii, T. (1970). *Wheat Inf. Serv. 31*, 1.

Genovesi, A.D., and Magill, C.W. (1979). *Crop Sci. 19*, 662.

George, L., and Rao, P.S. (1979). *Protoplasma 100*, 13.

Georgiev, G., and Chavdarov. L. (1974). *Genet. Sel. 7*, 404.

Gresshoff, P.M., and Doy, C.H. (1972a). *Aust. J. Biol. Sci. 25*, 259.

Gresshoff, P.M., and Doy, C.H. (1972b). *Planta 106*, 161.

Gu, S.R. (1979). *Acta Bot. Sci. 21*, 30.

Gu, S.R. (1981). *Acta Bot. Sci. 23*, 246.

Guha, S., and Maheshwari, S.C. (1964). *Nature (Lond.) 294*, 496.

Guha, S., and Maheshwari, S.C. (1966). *Nature (Lond.) 212*, 97.

Guha, S., and Maheshwari, S.C. (1967). *Phytomorphology 17*, 454.

Guha, S., Iyer, R.D., Gupta, N., and Swaminathan, M.S. (1970). *Curr. Sci. 39*, 174.

Guha-Mukhergel, S. (1973). *J. Exp. Bot. 24*, 139.

Guo, Z.S., Sun, A.C., Wang, Y.Y., Gui, Y.L., Gu, S.E., and Miao, S.H. (1978). *Acta Bot. Sin. 20*, 204.

Gupta, S.C., and Babbar, S.B. (1980). *Z. Pflanzenphysiol. 96*, 465.

Harn, C. (1972). *SABRAO News Lett. 4*, 27.

Hidaka, T., Yamada, Y., and Shichijo, T. (1979). *Japan J. Breed. 29*, 248.

Ho, T.F., Pao, J.J., Hu, Y., Wang, Y.F., and Chou, W.Y. (1978). *In* "Proceedings of Symposium on Anther Culture" (H. Hu, ed.), p.292. Science Press, Peking.

Horner, M., and Street, H.E. (1978). *Ann. Bot. 42*, 763.

Horner, M., and Mott, R.L. (1978). *Planta 156*, 158.

Hougas, R.W., and Gabert, A.C. (1964). *Crop Sci. 4*, 593.

Hsu, C., Yin, K.C., Chu, C.Y., Bi, F.Y., Wang, S.D., Liu, D.Y.. Wang, F.L., Chien, N.F., Chu, C.C., Wang, C.C., Sun, C.S. (1975). *Acta Genet. Sin. 2*, 294.

Hsu, T.H. (1978). *In* "Proceedings of Symposium on Anther Culture" p.277. Science Press, Peking.

Hu, C., Huang, S.C., Ho, C.P., Liang, H.C., Chung, C.C., and and Peng, L.P. (1978). *In* "Proceedings of Symposium on Plant Tissue Culture". p.87. Science Press, Peking.

Hu, H., Hsi, T.y., and Chia, S.E. (1978). *Acta Genet. Sin.* 5, 23.

Hu, H., Xi, Z.Y., Zhuang, J.J., Ouyang, J.W., Zheng, J.Z., Jia, S.E., Tia, X., Jing, J.K., and Zhou, S.M. (1979). *Acta Genetc. Sin.* 6, 322.

Huang, S.C., Hu, C., and Chuang, C.C. (1978). *In* "Proceedings of Symposium on Anther Culture" (H. Hu, ed.), p.279. Science Press, Peking.

Hughes, K.W., Bell, S.L., and Caponetti, J.D. (1975). *Can. J. Bot.* 53, 1442.

Irikura, Y., and Sakaguchi, S. (1972). *Potato Res.* 15, 170.

Iyer, R.D., and Raina, S.K. (1972). *Planta 104,* 146.

Jacobson, E., and Sopory, S.K. (1978). *Theor. Appl. Genet.* 52, 119.

Johansson, L., and Eriksson, T. (1977). *Physiol. Plant.* 40, 172.

Kameya, T., and Hinata, K. (1970). *Japan J. Breed.* 20, 82.

Kasha, K.J., and Kao, K.N. (1970). *Nature (Lond.)* 225, 874.

Kasperbauer, M.J., and Collins, G.B. (1971). *Crop Sci.* 12, 98.

Kaul, B.L., and Zutschi, U. (1971). *Indian J. Exp. Biol.* 9, 522.

Keller, W.A., Rajhathy, T., and Lacapra, J. (1975). *Can. J. Genet. Cytol.* 17, 655.

Keller, W.A., and Amstrong, K.C. (1977). *Can. J. Bot.* 55, 1383.

Keller, W.A., and Stringham, G.R. (1978). *In* "Frontiers of Plant Tissue Culture, 1978", (T.A. Thorpe, ed.). p.113. Univ. of Calgary, Calgar, Alberta.

Kermicle, J.L. (1969). *Science 166,* 1422.

Kimata, M., and Sakamoto, S. (1972). *Japan J. Genet.* 47, 61.

Kimber, G., and Riley, R. (1963). *Bot. Rev.* 29, 480.

Konar, R.N., and Singh, M.N. (1979). *Z. Pflanzenphysiol.* 95, 87.

Kohhar, T.S., Bhalla, P.S., and Sabharwal, P.S. (1970). *Can. J. Bot.* 49, 391.

Kohlenbach, H.W., and Geier, T. (1972). *Z. Pflanzenphysiol.* 67, 161.

Kohlenbach, H.W., and Wernicke, W. (1978). *Z. Pflanzenphysiol.* 86, 463.

Ku, M.K., Cheng, W.C., Kuo, L.C., Kuan, Y.L., An. H.P., and Huang, C.H. (1978). *In* "Proceedings of Symposium on Plant Tissue Culture". p.35. Science Press, Peking.

Kuo, J.S., Wang, Y.Y., Chien, N.F., Ku, S.J., Kung, M.L., and Hsu, H.C. (1973). *Acta Bot. Sin.* 15, 37.

La Rue, C.D. (1948). *Bull. Torrey Bot.* 75, 597.

Lu, C.H., Chang, H.F., and Liu, Y.H. (1978a). *In* "Proceedings of Symposium on Anther Culture" (H. Hu. ed.). p.195. Science Press, Peking.

Lu, C.H., Chang, H.F., and Lie, Y.H. (1978b). *In* "Proceedings
    of Symposium on Plant Tissue Culture". p.24. Science
    Press, Peking.
Malhotra, K., and Maheshwari, S.C. (1977). *Z. Pflanzenphysiol.*
    85, 177.
Maliga, P., Marton, L., and Sz-Breznovists, A. (1973a). *Plant*
    Sci. Lett. 1, 119.
Marton, L., and Maliga, P. (1975). *Plant Sci. Lett. 5*, 77.
Matsubayashi, M., and Kuranuki, K. (1975). *Sci. Rep. Fac.*
    Kobe Univ. 12, 173.
McComb, J.A. (1978). *In* "Proceedings of Symposium on Plant
    Tissue Culture". p.167. Science Press, Peking.
Miao, S.H., Kuo, C.S., Kwei, Y.L., Sun, A.T., Ku, S.Y., Lu,
    W.L., Wang, Y.Y. (1978). *In* "Proceedings of Symosium
    on Plant Tissue Culture". p.23. Science Press, Peking.
Mix, G., and Foroughi-Wehr, B. (1978). *Barley Genet. News-*
    letter 8, 77.
Montelongo-Escobedo, H., and Rowe, P.R. (1969). *Euphytica 18*,
    116.
Nakamura, A.R., Yamada, T., Kadotani, N., and Itagaki, R.
    (1974a). *In* "Haploids in Higher Plants: Advances and
    Potential" (K.J. Kasha, ed.). p. 277. Univ. of Guelph
    Press, Guelph, Ontario.
Nakamura, A., Yamada, T., Kadotani, N., Itagaki, R., and
    Oka, M. (1974b). *SABRAO 6*, 107.
Nakata, M., and Tanaka, M. (1968). *Japan J. Genet. 43*, 65.
Narayanaswamy, S., and Chandy, L.P. (1971). *Ann. Bot. 35*,
    535.
Niizeki, M., and Grant, W.F. (1971). *Can. J. Bot. 49*, 2041.
Niizeki, M., Kita, F., and Takahashi, M. (1968). *Japan J.*
    Breed. 24, 304.
Niizeki, H., and Oono, K. (1968). *Proc. Japan Acad. 44*, 554.
Niizeki, H., and Oono, K. (1971). *Colloq. Int. CNRS 193*, 251.
Nilson-Tillgren, T., and von Wettstein-Knowles, P. (1970).
    Nature (Lond.) 227, 1256.
Nitsch, C. (1974). *C.R. Acad. Sci. 278D*, 1031.
Nitsch, C., and Norreel, B. (1973). *C.R. Acad. Sci. 276D*,
    303.
Nitsch, J.P. (1969). *Phytomorphology 19*, 389.
Nitsch, J.P., and Nitsch, C. (1969). *Science 163*, 85.
Nitsch, J.P., Nitsch, C., and Hamon, S. (1969a). *C.R. Acad.*
    Sci. 269D, 1275.
Nitsch, J.P., Nitsch, C., and Poereau-Leroy, P. (1969b). *C.R.*
    Acad. Sci. 269D, 1650.
Nitsche, W., and Wenzel, G. (1977). *Z. Pflanzenzuchtg. 8*,
    132.
Nostog, K. (1965). *Amer. J. Bot. 52*, 993.

Novak, F.J. (1974). *Z. Pflanzenzuchtg. 72*, 46.

Oka, M., Namura, A., and Yamada, T. (1976). *Bull. d'Information Coresta Special.* p.114.

O'Mara, J.G. (1940). *Genetics 32*, 99.

Ouyang, T.W., Hu, J., Chuang, C.C., and Tseng, C.C. (1973). *Sci. Sin. 16*, 79.

Pan. J.L., and Gao, G.H. (1975). *Acta Bot. Sin. 17*, 161.

Pan, J.L., and Gao, G.H. (1978). *Acta Bot. Sin. 20*, 122.

Pelletier, G., Raquin, C., and Simon, G. (1972). *C.R. Acad. Sci. 274D*, 848.

Phillips, G.C., and Collins, G.B. (1977). *Tob. Int. 179*, 69.

Picard, E.J., and de Buyser, J. (1973). *C.R. Acad. Sci. 277D*, 1436.

Picard, E.J., and de Buyser, J. (1977). *Ann. Amelior, Plantes 27*, 483.

Radojevic, L. (1978). *Protoplasma 96*, 369.

Raghaven, V. (1978). *Amer. J. Bot. 65*, 984.

Raghaven, V. (1975). *Z. Pflanzenphysiol. 76*, 89.

Raina, S.K., and Iyer, R.D. (1973). *Z. Pflanzenzuchtg, 70*, 275.

Rangaswamy, N.S. (1977). *In* "Plant Cell, Tissue and Organ Culture" p. 412. Springer-Verlag, Berlin.

Raquin, C., and Pilet, V. (1972). *C.R. Acad. Sci. 274D*, 1019.

Rashid, A., and Street, H.E. (1973). *Planta 113*, 263.

San Noeum, L.H. (1979). *In* "Proc. Conf. Broadening Genet. Base Crops" p.327. Pudoc, Wageningen.

Sangwan, R.S., and Norreel, B. (1975a). *Naturwiss. 62*, 440.

Sangwan, R.S., and Norrell, B. (1975b). *Nature 257*, 222.

Sangwan-Norreel, B.S. (1977). *J. Exp. Bot 28*, 843.

Scowcroft, W.R. (1977). *Adv. Agronomy 29*, 39.

Sehgal, C.B. (1978). *Z. Pflanzenphyiol. 88*, 349.

Sharp, W.R., Dougall, D.K., and Paddock, E.F. (1971). *Bull. Torrey Bot. Club 98*, 219.

Sharp, W.R., Raskin, R.S., and Sommer, H.E. (1972). *Phyto-morphology 21*, 334.

Sinha, S., Roy, R.P., and Jha, K.K. (1979). *Can. J. Bot. 57*, 2524.

Sink, Jr K.C., and Padman, Bham, V. (1977). *Hort. Science 12*, 143.

Smith, H.H. (1974). *Polyploidy Induced Mutat. Plant Breed. Proc. Eucarpia/FAO/IAFA*, 1972.

Sopory, S.K. (1977). *Z. Pflanzenphysiol. 84*, 453.

Sopory, S.K., and Maheshware, S.C. (1976). *J. Exp. Bot. 27*, 58.

Sopory, S.K., Jacobson, E., and Wenzel, G. (1978). *Plant Sci. Lett. 12*, 47.

156   Chih-Ching Chu

Subrahmanyam, N.C., and Kasha, K.J. (1973). *Chromosoma 42*, 111.
Subrahmanyam, N.C., and Kasha, K.J. (1975). *Can. J. Genetc. Cytol. 17*, 573.
Sun, C.S., Wang, C.C., and Chu, C.C. (1973). *Acta Bot. Sin. 15*, 163.
Sun, C.S., Wang, C.C., and Chu, C.C. (1974). *Sci. Sin. 17*, 793.
Sun, C.S., Chu, C.C., and Wang, C.C. (1978a). *Acta Bot. Sin. 20*, 210.
Sun, C.S., Wu, S.C., Wang, C.C., and Chu, C.C. (1978b). *Theor. Appl. Genet. 55*, 193.
Sun, C.S., Chu, C.C., Wang, C.C., and Tigerstedt, P.M.A. (1980). *Acta Bot. Sin. 22*, 27.
Sunderland, N. (1970). *New Scient. 47*, 142.
Sunderland, N., and Wicks, F.M. (1971). *J. Exp. Bot. 22*, 213.
Sunderland, N. (1977). *In* "Plant Tissue and Cell Culture" (H.E. Street, ed.) p. 223. Univ. of California Press, Berkeley.
Sunderland, N., and Roberts, M. (1977). *Nature (Lond.) 270*, 236.
Sunderland, N. (1978). *In* "Proceedings of Symposium on Plant Tissue Culture" p.65. Science Press, Peking.
Symko, S. (1969). *Can. J. Genet. Cytol. 11*, 602.
Takahashi, H. (1973). *Japan J. Breed. 23*, 250.
Teng, L.P., and Kuo, Y.H. (1978). *In* "Proceedings of Symposium on Anther Culture" (H. Hu, ed.) p. 198. Science Press, Peking.
Thomas, E., and Wenzel, G. (1975a). *Naturwiss. 62*, 40.
Thomas, E., and Wenzel, W. (1975b). *Z. Pflanzenzuchtg. 74*, 77.
Thomas, E., Hoffmann, F., and Wenzel, G. (1975). *Z. Pflanzenzuchtg. 75*, 106.
Thomas, E., King, P.J., and Potrykus, I. (1979). *Z. Pflanzenzuchtg, 82*, 1.
Tsay, H.S., Tseng, M.T. (1979). *Bot. Bull. Acad. Sin. 20*, 117.
Tsun, C.Y. (1978). *In* "Proceedings of Symposium on Anther Culture" p.297. Science Press, Peking.
Vasil, I.K. (1980). *Inter. Rev. Cytol. Supplement 11A*, 195.
Vishino, A., Babbar, S.B., and Gupta, S.C. (1979). *Z. Pflanzenphysiol. 94*, 169.
Vyskot, B., and Novak, F.J. (1974). *Theor. Appl. Genet. 44*, 138.
Wakasa, K., and Watanabe, Y. (1979). *Japan J. Breed. 29*, 146.
Wang, C.C., Chu, C.C., Sun, C.S., Wu, S.H., Yin, K.C., and Hsu, C. (1973). *Sci. Sin. 16*, 218.

Wang, C.C., Sun, C.S., and Chu, C.C. (1974). *Acta Bot. Sin.* *16*, 43.

Wang, C.C., Chu, C.C., and Sun, C.S. (1975a). *Acta Bot. Sin.* *17*, 56.

Wang, C.C., Chu, C.C., Sun, C.S., Hsu, C., Yin, K.C., and Bi, F.Y. (1975b). *Acta Genet. Sin.* *2*, 72.

Wang, C.C., Sun, C.S., and Chu, C.C. (1977). *Acta Bot. Sin.* *19*, 190.

Wang, C.C., Chu, C.C., and Sun, C.S. (1980). *Acta Bot. Sin.* *22*, 316.

Wang, C.C., and Kuang, B.J. (1981). *Acta Bot. Sin. 23*, (in press).

Wang, P., and Chen, Y.R. (1980). *Acta Genet. Sin. 7*, 64.

Wang, Y.Y., Sun, C.S., Wang, C.C., and Chien, N.F. (1973). *Sci. Sin. 16*, 147.

Weatherhead, M.A., Burdon, J., and Henshaw, G.G. (1978). *Z. Pflanzenphysiol. 89*, 141.

Weatherland, M.A., and Henshaw, G.G. (1979). *Euphytica 28*, 765.

Wenzel, G., Hoffman, I., Potrykus, I., and Thomas, E. (1975). *Mol. Gen. Genet. 138*, 293.

Wenzel, G., Hoffman, I., and Thomas, E. (1976). *Theor. Appl. Genet. 48*, 205.

Wenzel, G., Hoffman, F., and Thomas, E. (1977a). *Theor. Appl. Genet. 51*, 81.

Wenzel, G., Hoffman, I., and Thomas, E. (1977b). *Z. Pflanzen-zuchtg. 78*, 149.

Wernicke, W., and Kohlenbach, H.W. (1975). *Z. Pflanzenphysiol. 77*, 89.

Wernicke, W., and Kohlenbach, H.W. (1976). *Z. Pflanzenphysiol. 79*, 189.

Wernicke, W., and Kohlenbach, H.W. (1977). *Z. Pflanzenphysiol. 81*, 330.

Widholm, J.M. (1977). *In* "Plant Tissue Culture and its Bio-technological Applications" (W. Barz, E. Reinhard, and M. H. Zenk, ed.) p. 112. Springer, Berlin.

Wilson, H.M. (1977). *Plant Sci. Lett. 9*, 233.

Wilson, H.M., Mix, G., and Foroughi-Wehr, B. (1978). *J. Exp. Bot. 29*, 227.

Woo, S.C., Mok, T., and Huang, J.Y. (1978). *Bot. Bull. Acad. Sin. 19*, 171.

Wu, J.L., Zhong, Q.L., Nong, F.H., and Chang, T.M. (1980). *Hereditas (Beijing) 2*, 23.

Yin, K.C., Hsu, C., Chu, C.Y., Bi, Y., Wang, S.T., Liu, T.Y., Chu, C.C., Wang, C.C., and Sun, C.S. (1976). *Sci. Sin. 19*, 227.

Yin, K.C., Li, S.C., Xu, Z., Chen, L., Zhu, Z.Y., and Bi, F.Y.
    (1980). *Kexue Tongbao 25*, 864.
Zamir, D., Jones, R.A., and Kedar, N. (1980). *Plant Sci.
    Lett. 17*, 353.
Zenkteler, M. (1971a). *Experientia 27*, 1087.
Zenkteler, M. (1971b). *Genet. Pol. 12*, 267.
Zenkteler, M. (1972). *Biol. Plant. 14*, 420.
Zenkteler, M. (1973). *Z. Pflanzenphysiol. 69*, 189.
Zenkteler, M., and Misiura, E. (1974). *Biochem. Physiol.
    Pflanz. 165*. 337.
Zhong, Z.X., Ren. Y.Y., and Dai, W.P. (1978). *Acta Bot. Sin.
    20*, 180.
Zhou, C.J., and Li, P.F. (1981). *Acta Bot. Sin. 23*, 79.
Zhou, C., and Yang, H.Y. (1980). *Acta Genet. Sin. 7*, 287.
Zhou, C., and Yang, H.Y. (1981). *Plant Sci. Lett. 20*, 231.
Zhu, X.Y., Wang, R.L., and Liang, Y. (1980). *Sci. Silval
    Sinicae 16*, 190.
Zhu, Z.C., and Wu, H.S. (1979). *Acta Genet. Sin. 6*, 181.

CHAPTER 8

SOMACLONAL VARIATION: A NEW OPTION
FOR PLANT IMPROVEMENT

W.R. Scowcroft
P.J. Larkin

CSIRO Division of Plant Industry
Canberra, A.C.T.

I.   INTRODUCTION

In the 1970's one hectare of arable land produced
sufficient food to support an average of 2.6 persons. By the
year 2000 that same hectare will have to support 4 persons
(The Global 2000 Technical Report, 1980). This will necess-
itate a sustained average annual rate of increase in world
food production of about 2.2 percent; a figure achieved in
only a few record years since the 1950's.

Much of the increase in food production will come from
more extensive use of energy intensive inputs such as fertil-
isers, pesticides, herbicides and irrigation. But, unfortunately, in many cases with an increasing rate of diminishing
return. Moreover the less developed countries are financially
less able to exploit such energy intensive inputs.

Difficulties in sustaining an adequate food producing
capacity are likely to be exacerbated by the actual or planned
launching in several countries of alcohol fuel programs which
use agricultural commodities as feed stock (Brown, 1980).

The implications of such prognostications is that crop
productivity must be consistently increased. In this the
plant breeder will, as he has done in the past, figure prom-
inently. But to be increasingly effective the plant breeder's
armory will need to be augmented by more efficient technology.

Several possible innovations derived from developments in
plant cell culture. These include the ability to apply cellular selection for recovering useful genetic variants, anther
culture to speed the attainment of homozygosity, somatic

PLANT IMPROVEMENT
AND SOMATIC CELL GENETICS

159

hybridization for recombining genomes of sexually incompatible species and more recently the possibility of specific gene addition or modification by recombinant DNA techniques (Scowcroft, 1977; Thomas *et al.*, 1979; Kado and Kleinhofs, 1980). These approaches have made only limited direct contribution to genotype improvement of commercial species to date.

II. SOMACLONAL VARIATION

As so often happens valuable scientific applications emerge from unexpected quarters. Since Steward (1958) first demonstrated that plants could be regenerated from cultured cells, the use of a tissue culture cycle for clonal propagation has been developed for a vast array of plant species including many of our important agricultural species (Thomas *et al.*, 1979; Tisserat *et al.*, 1980).

A tissue culture cycle involves the establishment of a more or less dedifferentiated cell or tissue culture under defined culture conditions, proliferation for a number of cell generations and the subsequent regeneration of plants. In other words one imposes a period of essentially dedifferentiated cell proliferation between an explant and the next plant generation. The ease and efficiency with which such manipulations can be made varies enormously from species to species.

Because a culture cycle was seen essentially as a method of cloning a particular genotype it became the accepted dictum that all plants arising from a tissue culture should be exact copies of the parental plant, despite the occurrence of phenotypic variants amongst regenerated plants. These were usually dismissed, somewhat embarrassingly, as "artefacts of tissue culture". Sometimes the variants were viewed as consequences of the recent exposure to exogenous phytohormones and sometimes they were labelled "epigenetic" events which somehow made them unworthy of further scientific scrutiny. Recent evidence has shown this judgement to be premature and possibly erroneous.

Skirvin and Janick (1976a) were the first to emphasis the importance of clonal variants ("calliclones") in genotype improvement of horticultural species. Recently, Shepard *et al.* (1980) have demonstrated extensive variability among plants (protoclones) regenerated from cultured leaf protoplasts of potato. In each case tissue culture *per se* appeared to be an unexpectedly rich and novel source of genetic variability. In investigating the extent and possible significance of this source of genetic variability Larkin and Scowcroft (1981b) suggested that "somaclonal variation" might

be an adequate general term for variation among plants
regenerated from cultured cells or tissue.

A. *Sugar Cane*

The first real developments to utilize somaclonal variation in plant improvement began in the experimental station
of the Hawaiian Sugar Planter's Association. Details of
early results published in annual reports can be found in
Heinz and Mee (1969,1971) and Heinz (1973) where variation in
sugar cane somaclones was observed in morphological, cyto-
genetic and isozyme traits.
Following these exciting indications of variability in
sugarcane, work was begun in Fiji in particular to seek
resistance to Fiji disease virus (Krishnamurthi and Tlaskal,
1974). As early as 1970 they began screening the somaclones
of a number of varieties for their reaction to Fiji disease
(a leafhopper transmitted virus) and Downy Mildew
(*Sclerospora sacchari*). In every case some somaclones were
identified with increased resistance to both Fiji disease and
Downy Mildew. Some examples of the variation generated
by a tissue culture cycle are illustrated in Figs. 1 and 2.
There appears to be a predominant shift toward increased
resistance for both diseases. This is apparent even when the
donor variety is already reasonably resistant (e.g. LF60-3879,
Fig. 1). Some of the Pindar somaclones resistant to both Fiji
disease and Downy Mildew have been tested for yield in Fiji
and independently in Australia. They did not show a reduced
sucrose yield compared to Pindar, indeed some showed a
slightly increased yield though this was not statistically
significant.
An independent series of (more than 4000) somaclones have
been produced in Hawaii to screen for Fiji disease resistance.
The first 735 of these which were derived from varieties with
high susceptibilities (> 7 on the 1-9 disease rating scale)
contained 18% with tolerant or resistant ratings of < 5
(Heinz, 1976). From a different parent, somaclones were also
found which vary greatly in their reaction to eyespot disease
(*Helminthosporium sacchari*) (Heinz, 1973; Heinz *et al.*,
1977). It had been intended to induce mutations with methyl-
methanesulfonate (MMS) or ionizing radiation during culture
but they found the background variation was just as high as in
the mutagenized cultures (Fig. 3).
In 1979 we began exploring the potential of somaclonal
variation to improve Australian sugar cane cultivars particularly with respect to eyespot disease resistance. All lines

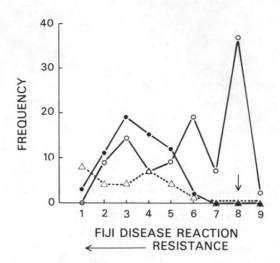

FIGURE 1.   The distribution of Fiji disease reaction of the
somaclones generated from three parental lines LF66-9601
(•——•), Pindar (o——o), and LF60-3879 (△——△).   All three
parental lines are highly susceptible (8) as indicated by the
arrow.   Derived from Krishnamurthi (1974).

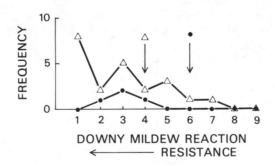

FIGURE 2.   The distribution of downy mildew reaction of soma-
clones from cultivars LF60-3879 (△——△) and LF51-124 (•——•).
The parental reactions are indicated by arrows.   Derived from
Krishnamurthi (1974).

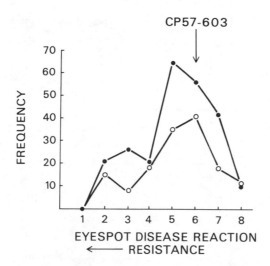

FIGURE 3. The distribution of the eyespot disease rating of somaclones from CP57-603 cultures with (●——●) or without (o——o) mutagenesis. The reaction of the parental line is indicated by the arrow. Derived from Heinz et al. (1977).

were initiated from the cultivar Q101 which is an agronomically valuable cultivar whose major fault is susceptibility to eyespot disease. Following prolonged culture (up to 15 mths) a large number of plants were regenerated and subsequently grown under glasshouse conditions.

To facilitate screening a large number of somaclones a leaf bioassay was developed to quantify the sensitivity of the leaves of a given plant to a standardised concentration of the fungal toxin responsible for leaf damage. The sensitity is measured as the initial rate of leakage of electrolytes from leaf discs briefly exposed to the toxin. This metric proved to be highly repeatable and consistent for a given cultivar (Larkin and Scowcroft, 1981a).

A potentially interesting and for our purposes a very useful sidelight emanated from the development of this bioassay. We found that co-culturing the fungus with suspensions of sugar cane cells resulted in an amplification of toxin production up to 4,000 fold greater than the level produced by pure fungal cultures.

The distribution of toxin sensitivity among these somaclones is depicted in Figure 4. Many of the somaclones

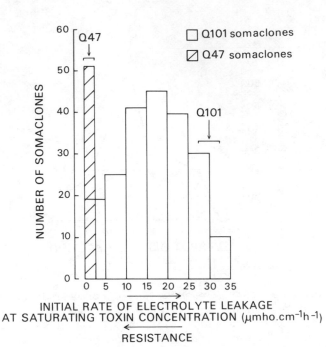

FIGURE 4. The distribution of reaction to eyespot toxin of Q101 somaclones, and Q47 somaclones. The parental reactions are as indicated by arrows.

proved to be resistant or essentially immune to the effects of the toxin ($< 5 \ \mu$mho.cm^{-1}.h^{-1}) and the mode of the distribution is significantly shifted to the resistant side of Q101. All 52 somaclones of Q47, a highly resistant cultivar, are also resistant. Most importantly these resistant somaclones retain their resistance through subsequent cane generations. Second and subsequent tissue culture cycle somaclones of resistant somaclones also tend to retain toxin insensitivity. However, there is some transgression of individual plants toward susceptibility among these later generation somaclones.
     Liu and Chen (1976,1978) in Taiwan have found significant variations amongst sugar cane somaclones from 8 varieties in characters such as cane yield, sugar yield, stalk number, length, diameter, volume, density and weight, percent fiber, auricle length, dewlap shape, hair group, and attitude of top leaf. Some of these somaclones in replicated randomized complete block field experiments showed significant improvements over the parental performance and over the performance

of Taiwan's major variety.   Liu and Chen have also recovered
resistant somaclones from susceptible donors for both Downy
Mildew and also culmicolous smut diseases.

These several independent reports of extensive somaclonal
variation in sugar cane strongly recommend that cell culture
become an integral part of sugar cane improvement. This
approach not only is useful in improving sugar cane for traditional purpose but might have a significant role in rapidly
adapting this crop species for newer roles such as a more
efficient feedstock for alcohol fermentation.

B.   *Tobacco*

Tobacco is the doyen of species where tissue culture is
concerned, particularly in respect of plant regeneration. It
might therefore be expected that somaclonal variation should
be manifested in this species. Indeed from as early as 1969
variation was observed among tobacco somaclones (Sacristan and
Melchers, 1969) and subsequently has been consistently
reported (Larkin and Scowcroft, 1981b). Several of these
studies, particularly those which observe somaclonal variation
among androgenically derived double haploids provide com-
pelling evidence for the importance of somaclonal variation.

A substantive demonstration of tobacco somaclonal
variation was provided by Burk and Matzinger (1976). Their
parental plant was the highly inbred variety, Coker 139, which
had gone through 15 generations of self-fertilisation since
its release as a pure line. Dihaploids were derived by anther-
culture from this highly inbred line. Five spontaneously
doubled haploids were included with 41 colchicine doubled
haploids as a control on the effect of colchicine. The selfed
progeny of these lines and Coker 139 were evaluated in a
randomized block field design with six replications. Each
block contained 20 plants. Significant variability between
the somaclonal lines was observed in all the characters
examined: yield, grade index, days to flowering, plant height,
leaf number, leaf length, leaf width, total alkaloids and
reducing sugar content. The dihaploid lines which had not
been exposed to colchicine showed as much variability as those
that had been. There was no significant variation within
lines. It seems very unlikely that there could be any
residual heterozygosity in their parental plant and yet the
dihaploid somaclones displayed as much variability as normally
associated with a segregating F_2 population from a cross
between 2 different cultivars.

Most recently Barbier and Dulieu (1980) have made a painstaking and detailed analysis of a qualitatively staggering
amount of variation among plants regenerated by cultured cells
of tobacco. From a hybrid, heterozygous for 2 loci (a_1, yg)
which affect chlorophyll synthesis (yellowish-green phenotype)
they found that the frequency of mutations (reversions to a
wild type allele or deletions) among 1,666 plants regenerated
was 3.5% and 3.6% at each locus respectively. This estimate
was obtained by genetically analysing 113 more or less green
variant plants. The authors acknowledge that this is an under-
estimate resulting from misclassification of variants as
parental types at the plantlet stage.

In their experiment they observed the frequency of variants regenerated directly from cotyledon explants (30 days in
culture), from cotyledon derived callus cultured in an
undifferentiated state for 135, 180 or 255 days from callus
initiation to plantlet formation and among plants regenerated
from cotyledon protoplasts (80 days from protoplast culture to
plants). They found that the frequency of mutations at the a_1
and yg loci was at a maximum in plants derived from protoplasts
and early callus cultures, and remained stable among plants
regenerated from older callus. For these they conclude that
the mutational event preexisted in the cotyledon cells either
as a functional mutation or as a lesion in the DNA of the
latent cotyledon cell which on dedifferentiation is repaired
and expressed following mitotic segregation. It would seem
difficult to distinguish between variation preexisting in the
cotyledon cells and that generated during early culture phases.

C.   Potato

Perhaps the most significant demonstration of the potential value of somaclonal variation was provided by Shepard *et
al*. (1980) in potato. In North America the 70 year old
variety Russet Burbank represents 39% of the potato crop. This,
along with most 'old' potato varieties, has been retained
because although it is is not particularly remarkable in any
specific character, it is a good allrounder. Nevertheless,
each year about 22% of the world potato crop is lost because
of disease. The sterility or very low fertility of many
cultivars discourages their use in breeding programs.

Shepard *et al*. (1980) argued that it might be simpler to
selectively improve a popular variety than to create a new
one. Screening over 1000 somaclones ("protoclones") produced
from leaf protoplasts of Russet Burbank, they found significant and stable variation in compactness of growth habit,

maturity date, tuber uniformity, tuber skin color, photoperiod requirements, and fruit production. Some characteristics, such as greater tuber uniformity and early onset of tuberisation, were agronomic improvements over the parent, Russet Burbank.

It is most significant that some somaclones were recovered which were resistant to disease pathogens. Five of 500 soma-clones were more resistant to *Alternaria solani* toxin than the parent and of these, four showed field resistance to early blight. About 2.5% (20 of 800) somaclones screened were resistant to late blight (*Phytophthora infestans*) some of which were resistant to multiple races of this pathogen (Matern *et al.*, 1978; Shepard *et al.*, 1980).

These variant somaclones have retained their phenotype through a number of vegetative generations. Sixty five selected somaclones have now been analysed in detail for vari-ability under field condition (Secor and Shepard, 1981). Among 35 characters analysed statistically significant variation was found for 22 characters. All clones differed from Russet Burbank in at least 1 characteristic and one somaclone differed for 17 characters. The modal class of 15 clones differed from Russet Burbank in 4 characters.

An interesting contrast was provided by Wenzel *et al.* (1979) who observed phenotypic variability from protoplast derived somaclones of potato dihaploids only after extended culture periods. It is not possible to rationalise these results with those obtained by Shepard *et al.* (1980) at this time. However, it is apparent that the Wenzel *et al.* (1979) dihaploid material can generate variation because it has been possible to select variant cell lines resistant to toxic culture filtrates of *Phytophthora infestans* (Behnke, 1980a) and *Fusarium oxysporum* (Behnke, 1980b) respectively. Plants derived from these respective toxic culture filtrates are also more resistant to the toxin than was the parental line.

D.   *Rice*

Among rice somaclones variants have been reported in characters such as number of tillers per plant, number of fertile tillers per plant, average panicle length, frequency of fertile seed, plant stature and flag leaf length (Nishi *et al.*, 1968; Henke *et al.*, 1978).

Oono (1978) has made a detailed analysis of some 800 somaclones derived from a homozygous line of a selfed doubled haploid. Chloroplast content, flowering date, plant height, fertility and morphology were examined in each of these soma-clonal derivatives and in each of two subsequent selfed

generations.   For these characters only 28% of the plants were
normal with respect to the parental phenotype.   There was wide
variation in seed fertility, plant height and heading date.
Chlorophyll deficiencies were seen in the second generation of
8.4% of the lines which is a comparable frequency to that
expected from X-ray and γ-irradiation.   Sectorial analysis of
plants derived from a single seed callus showed that at least
most of the variations were induced during culture and were
unlikely to pre-exist among   the 75 homozygous seeds used to
initiate the experiment.   In the second selfed generation after
somaclone regeneration some of the mutant characters were
segregating and some were fixed.   It was estimated that mutations affecting these five traits were induced in culture at a
rate of 0.03-0.07/cell/division.

E.   Other Species

Table 1 provides a list of other species in which variation has been observed among plants regenerated from cell
culture.   There are a number of notable feature about some of
these cases.   It has been shown that selection in maize cell
culture for resistance to southern corn leaf blight (*Drechlsera
maydis* Race T) could produce disease resistant plants
(Gengenbach *et al.*, 1977).   In confirming this, Brettel and
Ingram (1979) and Brettel *et al.* (1980) also showed that the
frequency of occurrence of resistant variants was high (35/60)
even among plants regenerated from cultures which had not been
exposed to the toxin.
    The use of somaclonal variation has already had an impact
on varietal production in horticulture.   Skirvin (1978) refers
to the release of new *Pelargonium* varieties selected from
among somaclonal variants.

III.   VARIATION AMONG INTERSPECIFIC HYBRID SOMACLONES

Techniques, such as embryo culture, are being used with
increasing frequency to facilitate the transfer of alien genes
into commercial cultivar gene pools (see James and Williams,
this volume).   Though interspecific hybrids may be produced,
it is often difficult to transgress desirable genetic information from an alien genome to that of a commercial cultivar
presumably because of lack of recombination between the dis-
parate genomes.   Recent evidence suggests that enhanced

Table 1. Additional species displaying somaclonal variation

Species	Variant characters	References
Oats	– plant ht. heading date, leaf striping, twin culms, awn morphology	Cummings et al., 1976
	– heteromorphic bivalents, ring chromosomes	McCoy et al., 1978
Maize	– abphyl syndrome, pollen fertility	Green, 1977
Barley	– plant ht, tillering, fertility	Deambrogio and Dale, 1980
Sorghum	– fertility, leaf morphology, growth habit	Gamborg et al., 1977
Onion	– bulb size and shape, clove no., aerial bulbil germination	Novak, 1980
Rape	– flowering time, glucosinolate content, growth habit	Hoffman, 1978; Wenzel, 1980
Pineapple	– spine and leaf-colour, wax secretion, foliage density, leaf width, leaf spines	Wakasa, 1979
Lettuce	– leaf wt, length, width, flatness and colour, bud number	Sibi, 1976
Pelargonium	– leaf shape, size and form, flower morphology, plant ht, fasciation, pubescence, anthocyanin pigmentation, essential oil composition	Skirvin, 1978; Tokumasu and Kato, 1979
Chrysanthemum	– flower colour, temperature for flower induction	Jung-Heiliger and Horn, 1980

transgression occurs among plants regenerated from cell cultures
of interspecific F₁ hybrids.

Ahloowalia (1976,1978) produced a triploid hybrid embryo
from a diploid *Lolium perenne* x tetraploid *L. multiflorum*
cross. This embryo was cultured to produce callus and more
than 2000 plants were regenerated from the callus over a
period of five years. The plants showed a wide variation in
leaf shape, size, floral development, growth vigour, survival
and perenniality. Some of the variants represented combinations
of parental characteristics which were agronomically valuable
and, according to the author, had not been observed in hybrids
which had not been through a tissue culture cycle. The seed
progeny also showed the variation. The first series of hybrid
rye-grass plants regenerated were triploid (2n = 21) but many
subsequent plants were 2n = 20 with one being 15 and one 18.
Meiotic chromosome behaviour suggested the occurrence of
reciprocal translocations, deletions and inversions.

For interspecific hybrids the length of the culture phase
prior to regeneration appears to influence the degree of
variation recovered. Kasperbauer *et al.* (1979) examined over
1200 somaclones derived from cultures of the F₁ between
Lolium multiflorum and *Festuca arundinaceae.* There was no
variation apparent among somaclones produced during the 1st
4 months of culture. After 40 weeks of culture however both
karyotypic and phenotypic variants occurred. Among the latter
there were plants which had no obvious karyotypic alterations.

Among *Hordeum* tissue culture regenerated plants, Orton
(1980a) and Orton and Steidl (1980) have observed polyploidy,
aneuploidy and chromosomal rearrangements. As expected such
abnormalities also occur in interspecific hybrids. Orton
(1980b) found enhanced multivalent formation among somaclones
derived from the F₁ of the sterile interspecific hybrid, *H.
vulgare* x *H. jubatum.* In the original hybrid there was
virtually no synapsis between the chromosomes from the two
different genomes. Exchange between these two genomes will be
made possible if breakage and reunion events occur during
culture and also by the relaxation of pairing suppression
between homoeologues in the meiotic behaviour of plants
derived from those cultures. There was much morphological
and isozyme variability amongst the somaclones derived from
cultures of this hybrid. A number of the regenerates were
haploid and generally of *H. vulgare* appearance. Hence *H.
jubatum* chromosomes had been eliminated. These *H. vulgare*-
type haploids had *H. vulgare* bands when analysed for esterase
and glutamate–oxaloacetate transaminase isozymes. Yet 2 of
the 5 examined also showed a few *H. jubatum* bands indicating
that some interspecific exchange had occurred prior to
chromosome elimination (Orton, 1980b).

IV. ORIGINS OF SOMACLONAL VARIATION

An understanding of the causes of somaclonal variation is important, first, to be able to enhance the level of variation when increased variability is the objective. Second, where the goal is simply to propagate a useful genotype, it would be desirable to be able to suppress the phenomenon.
We would judge that many different processes can contribute and that several of these operate simultaneously in the one culture. A discussion of some of these possible processes, of necessity, will be brief. Further detail is provided by Larkin and Scowcroft (1981b).

1. Gross karyotypic changes. These changes have been adequately documented both in cell cultures and in regenerated plants. However, karyotypic changes seem to be a reflection of the variation generating capacity of cell cultures rather than a cause of phenotypic variation. Shepard *et al.* (1980) reported that five protoclones with potentially increased tuber yield all had a normal karyotypes. Similarly, all of the sorghum somaclones had a normal karyotype even those with leaf and growth habit abnormalities and Skirvin and Janick (1976a) found that only a small proportion of the *Pelargonium* somaclones had ploidy changes. Burk and Chaplin (1980) also found normal chromosome number and meiotic behaviour in a random sample of their tobacco somaclones. Barbier and Dulieu (1980) observed some polyploidization of tobacco somaclones but this was independent of changes at loci affecting chlorophyll synthesis.

2. Cryptic chromosomal rearrangements. This could be a major mechanism to generate somaclonal variation. Chromosome breakage and reunion, multicentrics and translocations have been observed in plants derived from barley cultures (Foroughi-Wehr *et al.*, 1979; Orton, 1980a). The meiotic behaviour of chromosomes in plants regenerated from cell cultures of different species indicate the presence of reciprocal translocations, deletions, inversions, non-homologous translocations, acentric and centric fragments (Ahloowalia, 1976,1978; Cummings *et al.*, 1976; McCoy *et al.*, 1978; Green *et al.*, 1977; Novak, 1980). Each of these changes could have a genetic consequence affecting an individual phenotype.

Such chromosome rearrangements can result in loss of genetic material which may result in phenotypic variants. As well as affecting the gene in which the chromosomal break occurs, neighbouring genes, particularly those for which transcription may be coordinately regulated, will also be affected. If reunion or transposition to a different site occurs then distant gene functions may also be altered, i.e. position effect.

Cryptic changes not only result in the loss of genes and their functions but also the expression of genes which have hitherto been silent. For example, a rearrangement may delete or otherwise switch off a dominant allele allowing the recessive allele to affect the phenotype. Siminovitch (1976) refers to this phenomenon as culture-induced hemizygosity. If portions of the genome become effectively haploid many recessives will be expressed.

3. Somatic Crossing over and Sister Chromatid Exchange.
Estimates of somatic crossing over, based on the frequency of occurrence of twin spots in heterozygotes, range from 5.7 x 10^{-5} to 7.7 x 10^{-6} per spot capable mitosis, for soybean and tobacco, respectively (Evans and Padock, 1976). The frequency of somatic exchange is increased by applied agents such as X-irradiation and mitomycin C. A tissue culture environment may enhance the frequency of somatic crossing over and if a proportion of such exchanges were asymetric or between non-homologous chromosomes then genetic variants may be generated.

Asymetric sister chromatid exchange can lead to deletions and duplications which in somatic cells would segregate in subsequent mitoses. The frequency of sister chromatid exchange in plants is relatively high (e.g. 20.6 per cell per division in barley) (Schubert *et al.*, 1980). Sister chromatid exchange in eucaryotes has also been implicated in explaining rDNA inheritance patterns in Drosophila (Tartof, 1973) and yeast (Petes, 1980).

4. Transposable elements. Transposable genetic elements are stretches of DNA which can move from one locus in the genome to another independently of extensive sequence homology (Calos and Miller, 1980). The excision and reinsertion of the genetic element can directly affect the expression of neighbouring structural genes. Moreover imprecise excision of bacterial transposable elements may generate rearrangements (deletions, inversions) of adjacent chromosomal sequences.

In eucaryotes genetic evidence suggests that certain unstable mutants may be explained by transposable elements.

Controlling elements in maize (McClintock, 1956) have many
genetic properties which are analogous to transposable elements
in bacteria (Peterson, 1970; Fincham and Sastry, 1974;
Starlinger, 1980).   Other mutable loci have been described in
plants such as *Antirrhinum*, tobacco hybrids and soybean (see
Fincham and Sastry, 1974; Sand, 1976), tomato (Hagemann,
1958) and *Petunia* (Potrykus, 1970). Transpositional events in
mitochondrial DNA have also been implicated in the spontaneous
reversion to fertility of S male sterile cytoplasm in maize
(Levings *et al.*, 1980). To date there is no proof of a causal
relationship between genetic instability and the physical
excision and/or insertion of DNA sequences.

There is physical evidence for the mobility of certain
repetitive DNA in yeast (Cameron *et al.*, 1979) and Drosophila
cell lines (Finnegan, 1978). The changing position of such
DNA in yeast has recently been directly correlated with
mutational events at the his 4 locus in yeast (Roeder and Fink,
1980). Moreover, his^+ revertants presumably resulting from
the excision of a DNA element may contain chromosomal
aberrations such as deletions, translocations, transpositions
or inversions.

It is conceivable that transpositional events could play
a role in somaclonal variation. A tissue culture environment
may be conducive to sequence transposition. Such high mut-
ability and adaptiveness in somatic tissues has been referred
to as "somatic Darwinism" (Weill and Reynaud, 1980). In
plants there is the opportunity to incorporate such putatively
generated variants into germ line cells.

5. *Gene amplification and diminution.* A cell can
increase or decrease the quantity of a specific gene product
simply by differential gene amplification or diminution,
respectively. Indeed any gene for which there is no modu-
lation of expression might be readily amplified or diminished
under an appropriate selection regime. For example, selection
for resistance to methotrexate, which inhibits dihydrofolate
reductase activity, can led to a 200 fold amplification of
the gene coding for this enzyme (Schimke *et al.*, 1977).
Similarly, the gene complex which includes the coding sequence
for aspartate transcarbamylase is amplified by up to 190 fold
when hamster cells are selected for resistance to a transition
state analogue of aspartate transcarbamylase (Wahl *et al.*,
1979).

Such processes also seem to occur in plants. In particular ribosomal RNA gene amplification and depletion has been
documented in wheat, rye, hyacinth, maize, *Vicia*, melon and
tobacco (Flavell, 1975; Siegel, 1975). DNA of flax is known

to change in response to environmental pressure (Cullis, 1973; Cullis and Goldsborough, 1980). Reassociation kinetics showed that one plant form possessed a class of moderately repeated DNA not present in the other forms. Soybean cell cultures adapted to prolonged growth on maltose had lost one third of their ribosomal genes. This diminution can be reversed by readaption of the cell line to sucrose media provided the maltose adaptation was not prolonged.

If such amplification or depletion of DNA sequence can occur in plant cell cultures it may account at least in part for somaclonal variation. Though not proven, several cases of selection in plant cell cultures are consistent with gene amplification. Selection for resistance to high levels of *Drechslera maydis* Race T Race T toxin in toxin sensitive (T cytoplasm) maize cultures resulted from serial selection at progressively higher toxin levels (Gengenbach and Green, 1975). Similarly Nabors *et al*. (1980) required 11 successive selection cycles to obtain high level resistance to salinity in tobacco cultures.

V.   CONCLUSION

There is a growing awareness that for those species which are amenable to cell culture, somaclonal variation may provide an important source of variability for plant improvement. We have argued that somaclonal variation is widespread and is found not only in asexually propagated species but also in seed propagated, self-fertilizing species. Although the most extensive examples to date have been in non-diploids (sugar cane, potato and tobacco) it has been observed in diploid species also. The spectrum of characters affected can be diverse and the frequency of variant occurrence comparatively high.

The value of somaclonal variation to plant improvement depends on the ease with which given plant species can be cultured and plants regenerated from cell lines. Recent research is enhancing this capacity in previously recalcitrant crop plants such as wheat (Shimada and Yamada, 1979), rice (Oono, 1978) and important legumes (Bingham *et al*., 1975; Beach and Smith, 1979; Phillips and Collins, 1979; Kao and Michayluk, 1980; dos Santos *et al*., 1980) possibly including soybean (Saka *et al*., 1980).

Somaclonal variation may find its greatest application for plant improvement in concert with selection for desirable mutations at the cellular level. It should no longer be surprising that in the recovery of cell culture mutants the frequencies have been high and mutagenic treatments often

failed to noticeably enhance the frequency. Cellular
selection is conceivable for the recovery of variants
resistant to antimetabolites such as amino acid analogues,
antibiotic drugs, pathotoxins, herbicides and physiological
stress (see Maliga, 1978; Thomas *et al.*, 1979; Brettell and
Ingram, 1979). Many agronomically important attributes are
known or suspected of having a cellular basis. These include
resistance to host-specific toxins such as those found in some
of the *Drechslera, Pseudomonas* and *Alternaria* pathogens,
tolerance of adverse soils (salinity, metal toxicity),
herbicide tolerance, tolerance to temperature stresses and
waterlogging. Somaclonal variation, cellular selection and
early rapid screening of regenerants collectively provide a
powerful option for plant improvement.

REFERENCES

Ahloowalia, B.S. (1976). *In* "Current Chromosome Research"
    (eds. K. Jones and P.E. Brandham), p.115. Elsevier/North
    Holland Biomedical Press, Amsterdam.
Ahloowalia, B.S. (1978). Fourth Intl. Cong. Plant Tissue
    Cell Culture, (abst.) p.162. Calgary, Canada.
Barbier, M., and Dulieu, H.L. (1980). *Ann. Amélior. Plantes*
    30, 321.
Beach, K.H., and Smith, R.R. (1979). *Plant Sci. Lett. 16*,
    231.
Behnke, M. (1980a). *Theor. Appl. Genet. 56*, 151.
Behnke, M. (1980b). *Z. Pflanzemzuchtg 85*, 254.
Bingham E.T., Hurley L.V., Kaatz D.M., and Saunders J.W.
    (1975). *Crop Sci. 15*, 719.
Brettell, R.I.S., and Ingram, D.S. (1979). *Biol. Rev. 54*,
    329.
Brettell, R.I.S., Thomas, E., and Ingram D.S. (1980). *Theor.*
    Appl. Genet. 58, 55.
Brown, L.R. (1980). "Food or Fuel: New competition for the
    world's cropland". Worldwatch Institute, Washington.
Burk, L.G., and Chaplin, J.F. (1980). *Crop Sci. 20*, 334.
Burk, L.G., and Matzinger, D.F. (1976). *J. Hered. 67*, 381.
Calos, M.P., and Miller, J.H. (1980). *Cell 20*, 579.
Cameron, J.R., Loh, E.Y., and Davis, R.W. (1979). *Cell 16*,
    739.
Cullis, C.A. (1973). *Nature 243*, 515.
Cullis, C.A., and Goldsborough, P.B. (1980). *In* "The Plant
    Genome" (eds. D.R. Davies, D.A. Hopwood), p.91. The
    John Innes Charity, Norwich.

Cummings, D.P., Green, C.E., and Stuthman, D.D. (1976). *Crop Sci. 16*, 465.

Deambrogio, E., and Dale, P.J. (1980). *Cereal Res. Comm. 8*, 417.

dos Santos, A.V.P., Outka, D.E., Cocking, E.C., and Davey, M.R. (1980). *Z. Pflanzenphysiol. 99*, 261.

Evans D.A., and Paddock E.F. (1976). *Can. J. Genet. Cytol. 18*, 57.

Evans, D.A., Sharp, W.R., and Flick, C.E. (1980). *Hort. Reviews 2*, 214.

Fincham, J.R.S., and Sastry, G.R.K. (1974). *Ann. Rev. Genet. 8*, 15.

Finnegan, D.J., Rubin, G.M., Young, M.W., and Hogness, D.S. (1978). *Cold Spring Harbor Symp. Quant. Biol. 42*, 1053.

Flavell, R.B. (1975). *In* "Modification of the Information Content of Plant Cells" (R. Markham *et al.* eds.) p.53. North Holland, Amsterdam.

Foroughi-Wehr, B., Mix, G., and Friedt, W. (1979). *Barley Genet. Newslett. 9*, 20.

Gamborg, O.L., Shyluk, J.P., Brar, D.S., and Constabel, F. (1977). *Plant Sci Letts. 10*, 67.

Gengenbach, B.G., and Green, C.E. (1975). *Crop Sci. 15*, 645.

Gengenbach, B.G., Green, C.E., and Donovan, C.M. (1977). *Proc. Nat. Acad. Sci. (Wash.) 74*, 5113.

Green, G.E. (1977). *Hort. Sci. 12*, 7.

Green, C.E., Phillips, R.L., and Wang, A.S. (1977). *Maize Genet. Newslett. 51*, 53.

Hagemann, R. (1958). *Z. Verebungs. 89*, 587.

Heinz, D.J. (1973). *In* "Induced Mutations in Vegetatively Propagated Plants", p.53. International Atomic Energy Agency, Vienna.

Heinz, D.J. (1976). Annl. Rep. Hawaiian Sugar Planters Assoc. Expt. Station p.9.

Heinz, D.J., and Mee, G.W.P. (1969). *Crop Sci. 9*, 346.

Heinz, D.J., and Mee, G.W.P. (1971). *Am. J. Bot. 58*, 257.

Heinz, D.J., Krishnamurthi, M., Nickell, L.G., and Maretzki, A. (1977). *In* "Applied and Fundamental Aspects of Plant Cell, Tissue and Organ Culture". (Reinert, J. and Bajaj, Y.P.S. eds.) p.3. Springer-Verlag, Berlin.

Henke, R.R., Mansur, M.A., and Constantin, M.J. (1978). *Physiol. Plant 44*, 11.

Hoffmann, F. (1978). *In* "Production of Natural Compounds by Cell Culture Methods". (A.W. Alfermann and E. Reinhard), p.319. Gesellschaft für Strahlen – und Umweltforschung, München.

Jung-Heiliger, H., and Horn, W. (1980). *Z. Pflanzezuchtg. 85*, 185.

Kado, C.I., and Kleinhofs, A. (1980). *Intern. Rev. Cytol. Suppl. 11B*, 47.

Kao, K.W., and Michayluk, M.R. (1980). *Z. Pflanzenphysiol. 96*, 135.

Kasperbauer, M.J., Buckner, R.C., and Bush, L.P. (1979). *Crop Sci. 19*, 457.

Krishnamurthi, M. (1974). *Sugarcane Breeder's Newsletter 35*, 24.

Krishnamurthi, M., and Tlaskal, J. (1974). *Proc. Int. Soc. Sugar Cane Technol. 15*, 130.

Larkin, P.J., and Scowcroft, W.R. (1981a). *Plant Physiol. 67*, 408.

Larkin, P.J., and Scowcroft, W.R. (1981b). *Theor. Appl. Genet. 60*, 197.

Levings, C.S. III, Kim, B.D., Pring, D.R., Conde, M.F., Mans, R.J., Laughnan, J.R., and Gabay-Laughran, S.J. (1980). *Science 209*, 1021.

Liu, M-C., and Chen, W-H. (1976). *Euphytica 25*, 393.

Liu, M-C., and Chen, W-H. (1978). *Euphytica 27*, 273.

McClintock, B. (1956). *Cold Spring Harbor Symp. Quant. Biol. 21*, 197.

McCoy, T.J., Phillips, R.L., and Cummings, D.P. (1978). p.101, (abst.) Fourth Intl. Cong. Plant Tissue Cell Culutre, Calgary, Canada.

Maliga, P. (1978). *In* "Frontiers of Plant Tissue Culture 1978" (T.A. Thorpe, ed.) p.381. Int. Ass. Plant Tissue Culture, Calgary.

Matern, U., Strobel, G., and Shepard, J. (1978). *Proc. Nat. Acad. Sci. (Wash.) 75*, 4935.

Nabors, M.W., Gibbs, S.E., Bernstein, C.S., and Meis, M.E. (1980). *Z. Pflanzenphysiol. 97*, 13.

Nishi, T., Yamada, Y., and Takahashi, E. (1968). *Nature 219*, 508.

Novák, F.J. (1980). *Z. Pflanzenzüchtg. 84*, 250.

Oono, K. (1978). *Trop. Agric. Res. Series 11*, 109.

Orton, T.J. (1980a). *Theor. Appl. Genet. 56*, 101.

Orton, T.J. (1980b). *J. Heredity 71*, 780.

Orton, T.J., and Steidl, R.P. (1980). *Theor. Appl. Genet. 57*, 89.

Peterson, P.A. (1970). *Genetica 41*, 33.

Petes, T.D. (1980). *Cell 19*, 765.

Phillips, G.C., and Collins, G.B. (1979). *Crop Sci. 19*, 59.

Potrykus, I. (1970). *Z. Pflanzenzüchtg. 63*, 24.

Roeder, G.S., and Fink, G.R. (1980). *Cell 21*, 239.

Sacristan, M.D., and Melchers, G. (1969). *Molec. gen. Genet. 105*, 317.

Saka, H., Voqui-Dinh, T.N., and Chang, T.Y. (1980). *Plant Sci. Letts. 19*, 193.

Sand, S. (1976). *Genetics 83*, 719.

Schimke, R.T., Alt, F.W., Kellems, R.E., Kaufman, N., and
    Bertino, J.R. (1977). *Cold Spring Harbor Symp. Quant.
    Biol. 42*, 649.

Schubert, I., Künzel, G., Bretschneider, H., and Rieger, R.
    (1980). *Theor. Appl. Genet. 56*, 1.

Scowcroft, W.R. (1977). *Adv. Agronomy 29*, 39.

Secor, G.A., and Shepard, J.F. (1981). *Crop Sci. 21*, 102.

Shepard, J.F., Bidney, D., and Shahin E. (1980). *Science 28*,
    17.

Shimada, T., and Yamada, Y. (1979). *Jap. J. Genet. 54*, 379.

Sibi, M. (1976). *Ann. Amélior. Plantes 26*, 523.

Siegel, A. (1975). *In* "Modification of the Information
    Content of Plant Cells" (R. Markham *et al.* eds.) p.15.
    North Holland, Amsterdam.

Siminovitch, L. (1976). *Cell 17*, 1.

Skirvin, R.M. (1978). *Euphytica 27*, 241.

Skirvin, R.M., and Janick, J. (1976a). *J. Amer. Soc. Hort.
    Sci. 101*, 281.

Starlinger, P. (1980). *In* "Genome Organization and Expression
    in Plants" (C.J. Leaver ed.) p.537. Plenum Press, New
    York.

Steward, F.C. (1958). *Amer. J. Bot. 15*, 709.

The Global 2000 Report to the President. Entering the Twenty-
    First Century, vol. 2, Technical Report, Washington:
    Government Printing Office 1980.

Tartof, K.D. (1973). *Cold Spring Harbor Symp. Quant. Biol.
    38*, 491.

Thomas, E., King, P.J., and Potrykus, I. (1979). *Z.
    Pflanzenzuchtg 82*, 1.

Tisserat, B., Esan, E.B., and Murashige, T. (1979). *Hort.
    Reviews 1*, 1.

Tokumasu, S., and Kato, M. (1979). *Euphytica 28*, 329.

Wahl, G.M., Padgett, R.A., and Stark, G.R. (1979). *J. Biol.
    Chem. 254*, 8679.

Wakasa, K. (1979). *Jap. J. Breeding 29*, 13.

Weill, J.C., and Reynaud, C.A. (1980). *Biosystems 12*, 23.

Wenzel, G. (1980). *In* "The Plant Genome" (D.R. Davies and
    D.A. Hopwood eds.) p.185. The John Innes Charity,
    Norwich.

Wenzel, G., Schieder, O., Przewozny, T., Sopory, S.K., and
    Melchers, G. (1979). *Theor. Appl. Genet. 55*, 49.

CHAPTER 9

PLANT CELL CULTURE AND SOMATIC CELL GENETICS OF CEREALS AND GRASSES[1]

INDRA K. VASIL

Department of Botany
University of Florida
Gainesville, Florida

The successful regeneration of plants from single cells and protoplasts has caused a considerable upsurge of interest and activity in the possible combined uses of plant cell culture and modern molecular genetics techniques in the improvement of crop plants. Such discussion, and the inevitable speculation, has been buttressed by the development of the novel techniques of protoplast fusion and somatic hybridization, production of haploid plants, and the genetic transformation of plant cells by the incorporation of cell organelles, nuclei, DNA, plasmids, etc. (I.K. Vasil, 1980a, b), accompanied more recently by some of the most fascinating studies of genome organization and expression in plants resulting from the application of the techniques of nucleic acid hybridization, DNA cloning, DNA sequencing, and restriction endonuclease mapping (Leaver, 1980).

The serious political, economic and social ramifications of the ever increasing world population, and the consequent shortages of food, particularly proteins in the developing and most populous countries, have helped to focus the attention of scientists, foundations, governments,

[1]*Based on an article published in Plant Molecular Biology Association Newsletter (2, 9-23. 1981).*

international organizations, and the business
community, to the possible application of the so-
called "genetic engineering" techniques to the
improvement of crop plants. The treatises cited
above not only provide ample evidence of the
vigour, novelty and the future prospects of cell
and tissue culture and somatic cell genetics of
plants, but also indicate that many of the
predictions and expectations forecast by scientists
and laymen alike are overinflated, premature, and
unrealistic. Thus, it is my firm belief that the
developing techniques of cell culture and somatic
cell genetics will become an important adjunct to,
and will supplement, the conventional methods of
crop improvement in the years to come, quite
possibly before the end of this century, but will
not replace the elegant and sophisticated systems
evolved in nature over millions of years.

Readers of contemporary plant cell and tissue
culture literature can certainly not fail to notice
that the introductory sentences in most of the
published material uniformly imply that the results
of such research will help to produce improved
plants. However, an objective observer will soon
discover that such predictions, and hopes, are
based almost exclusively on work with model plant
systems like tobacco, carrot, etc. There is not
much advantage to be gained by the genetic
improvement of these plants, and their use as model
experimental systems will have been entirely in
vain if the knowledge gained from their use can not
be successfully adapted for important crop plants
like the cereals and the legumes which provide the
bulk of the nutrition, directly or indirectly, for
man. Some of these model systems have been in use
from the early 1930's when Gautheret, Nobecourt and
White pioneered the techniques of plant cell and
tissue culture, and have undoubtedly contributed
immensely to this field of experimental botany as
we know it today. Although further basic work on
model systems is important and should continue in
the future, I believe the time has come to begin to
apply the knowledge gained from the use of model
species to crop plants, and indeed to develop
similar model species for legumes and cereals. For
legumes, *Medicago sativa* and *Trifolium*, in which
regeneration from callus, cell suspension cultures
and protoplasts has been achieved (Saunders and

Bingham, 1972; McCoy and Bingham, 1977; Kao and
Michayluk, 1980; Gresshoff, 1980; Phillips and
Collins, 1980), can serve as useful models. For
vegetatively propagated tuber crops, the pioneering
and elegant work of Shepard and his colleagues
(Shepard et al., 1980) would serve as an excellent
example. For cereals, *Panicum maximum* (Lu and
Vasil, 1981a, b, 1982; Lu et al., 1981), *Pennisetum
americanum* (V. Vasil and Vasil, 1980, 1981a, b,
1982a, b) and *Zea mays* (Green and Phillips, 1975;
Gengenbach et al., 1977; Potrykus et al., 1979;
Hibberd et al., 1980; Hibberd and Green, 1982; Lu
et al., 1982) are the obvious choices.

Owing largely to my personal research interests
and objectives, this discussion is limited to the
cell culture and somatic cell genetics of cereals
and grasses.

The techniques of cell culture and somatic cell
genetics that must be successfully developed for
cereals and grasses are: (1) Rapid clonal
propagation; (2) Regeneration from single cells and
protoplasts; (3) Androgenetic haploids; (4)
Mutant/variant cell lines and plant regeneration;
(5) Somatic hybridization; (6) Transformation; and
(7) Molecular biology.

I. CLONAL PROPAGATION

All of the major species of cereals and grasses
can now be clonally propagated in vitro (I. K.
Vasil and Vasil, 1980a). However, several
important problems still must be studied and
resolved. For example, plant regeneration is often
sporadic and transient, and only a few genotypes of
each species respond favorably to morphogenetic
stimuli in vitro and tend to rapidly lose the
potential for regeneration after a few subcultures. Furthermore, in most instances plants
are formed in vitro through the development of
axillary or adventitious buds (Rangan, 1974;
Dunstan et al., 1979; Nakano and Maeda, 1979;
Shimada and Yamada, 1979; Springer et al., 1979).
The new shoots arise through a process of "micro-
tillering" (Dunstan et al., 1979; Thomas et al.,
1979). The organization of shoot meristems is
known to be multicellular in origin (Crooks, 1933;
I. K. Vasil and Hildebrandt, 1966; Steffensen,

1968; Coe and Neuffer, 1978). Plants of
multicellular origin can not always be expected to
be genetically uniform, and may indeed be chimeras
(Sacristan and Melchers, 1969; Ogura, 1976; Sree
Ramulu *et al.*, 1976; Bennici and D'Amato, 1978; Mix
et al., 1978) and unsuitable for mutation breeding,
genetic analyses, etc.

The phenomenon of somatic embryogenesis in the
Gramineae has been discovered only recently.
Embryogenic tissue cultures have been obtained from
cultured immature embryos, young inflorescences and
leaves, and protoplasts of many species (Dale,
1980; V. Vasil and Vasil, 1980, 1981a, b, 1982a, b;
Wernicke and Brettell, 1980; Brettell *et al.*,
1980b; Haydu and Vasil, 1981; Lu and Vasil, 1981a,
b, 1982; Lu *et al.*, 1981, 1982; Ozias-Akins and
Vasil, 1982; Wang and Vasil, 1982). It is clear
that somatic embryogenesis is of common occurrence,
or can be induced in tissue cultures of most of the
species of cereals and grasses (Table I). The
extent of embryogenesis, and the degree of normalcy
of the embryoids formed, vary from species to
species (Table II).

A reexamination of the published reports of
plant regeneration in tissue cultures of the
Gramineae indicates that embryogenic tissue
cultures were indeed formed in many cases.
Unfortunately, these were either not recognized, or
were misinterpreted as shoot forming cultures
because of the precocious germination of the
embryoids leading to the development of leafy
scutella followed by shoot morphogenesis. In other
instances the initially slowly growing compact
embryogenic tissues were either deliberately
discarded in favour of the faster growing but often
friable nonembryogenic tissues, or were slowly
diluted out during subculture by experimental
conditions which favored the growth of nonembryogenic cells. Our experience has shown that
the early recognition and physical separation of
the embryogenic cells is critical in establishing
stable embryogenic tissue cultures.

Table I. Somatic Embryogenesis in Tissue cultures of Cereals and Grasses

Species	Source of explant	Type of cell culture	References
Lolium multiflorum	immature embryo, inflorescence	callus	Dale (1980) Dale et al. (1981)
Panicum maximum	immature embryo, inflorescence, leaf	callus, cell suspension, protoplast	Lu and Vasil (1981a, b, 1982), Lu et al. (1981)
Pennisetum americanum	immature embryo, inflorescence, leaf	callus, cell suspension, protoplast	V. Vasil and Vasil (1980, 1981a, b, 1982a, b), Rangan and Vasil (unpublished)
Pennisetum americanum x P. purpureum	inflorescence	callus	V. Vasil and Vasil (1981a)
Pennisetum purpureum	leaf, anther, inflorescence	callus	Haydu and Vasil (1981), Wang and Vasil (1982)
Saccharum officinarum	leaf	callus, cell suspension	Ho and Vasil (unpublished)
Sorghum bicolor	immature embryo, inflorescence, leaf	callus	Thomas et al. (1977), Wernicke and Brettell (1980), Brettell et al. (1980b)
Sorghum arundinaceum	inflorescence	callus	Boyes and Vasil (unpublished)
Triticum aestivum	immature embryo	callus	Ozias-Akins and Vasil (1982)
Zea mays	immature embryo	callus, cell suspension, protoplast	Lu et al. (1982), Lu and Vasil (unpublished)

Table II. Degree and Extent of Somatic Embryogenesis Observed in Tissue Cultures of Grasses. A High Score Also Indicates That Well Organized and Typical Embryoids are Formed.[a]

SOMATIC EMBRYOGENESIS

Maximum 10	*Panicum maximum*
	Pennisetum americanum
	P. americanum x *P. purpureum*
9	*P. purpureum*
	Sorghum arundinaceum
7	*Zea mays*
5	*Saccharum officinarum*
	Sorghum bicolor
3	*Triticum aestivum*
None 0	*?*

[a]*Based on observations in the author's laboratory.*

In many species the best results are obtained when immature embryos are cultured. This often necessitates the selection of experimental material and the initiation of cultures before its genetic makeup can be determined. It would be of advantage, therefore, to be able to use explants from mature plants for culture and in this respect the recent success in the regeneration of plants from leaf (Wernicke and Brettell, 1980; Haydu and Vasil, 1981; Lu and Vasil 1981b) and inflorescence tissues (V. Vasil and Vasil, 1981a; Lu and Vasil, 1982; Brettell *et al.*, 1980b) is noteworthy.

II. REGENERATION FROM SINGLE CELLS AND PROTOPLASTS

Regeneration of plants from single cells is perhaps the most important and the basic pre-requisite that must be fulfilled before the techniques of somatic hybridization, somatic cell

genetics and "genetic engineering" can be usefully
employed for plant improvement. Development of
plants from protoplasts, which are single cells is,
therefore, considered a key step in all such
attempts.

Although plants have been developed from
protoplasts of many species, especially those
belonging to the Solanaceae and Cruciferae (I. K.
Vasil and Vasil, 1980b), protoplasts of
graminaceous species have proved to be extremely
recalcitrant and difficult to culture (Potrykus,
1980). Nevertheless, cell colonies and callus
tissues have been obtained from protoplasts of
Hordeum vulgare (Koblitz, 1976), *Oryza sativa*
(Deka and Sen, 1976; Cai *et al.*, 1978), *Pennisetum
americanum* (V. Vasil and Vasil, 1979), *Sorghum
bicolor* (Brar *et al.*, 1980), *Triticum monococcum*
(Nemet and Dudits, 1977) and *Zea mays* (Potrykus *et
al.*, 1979). In each of these cases protoplasts
were isolated from established cell cultures. In
none of the examples cited above, organized growth
into embryoids, shoots or plants was possible.

Considerable interest and attention has,
therefore, been attached to the report by V. Vasil
and Vasil (1980) in which they induced somatic
embryogenesis and plantlet formation in protoplast-
derived cell colonies of *Pennisetum americanum*.
They obtained protoplasts from an embryogenic
suspension culture which was itself derived from an
embryogenic callus tissue of immature embryo
origin. Such cultures contain clearly distinguishable nonembryogenic and embryogenic
cells. Protoplasts derived from only the latter
were shown to be capable of division and further
organized development.

The isolation of the embryogenic cell line and
of embryogenic protoplasts proved to be the
critical factor in the success with *Pennisetum
americanum*. Since the publication of this report,
there has been renewed activity in several
laboratories interested in cereal protoplast
culture to isolate morphogenic and/or embryogenic
cell suspension cultures. Such suspension cultures
have now also been established in *Panicum maximum*
(Lu and Vasil, 1981b) *Panicum purpureum* (Wang and
Vasil, unpublished), and *Zea mays* (Lu *et al.*, 1982;
Lu and Vasil, unpublished).

Embryogenic cell suspension cultures derived from immature embryos and young inflorescences of *Pennisetum americanum* (V. Vasil and Vasil, 1981b), 1982) and *Panicum maximum* (Lu and Vasil, 1981a) can be experimentally manipulated to establish cultures that are composed predominantly of embryogenic cells. Protoplasts isolated from such suspension cultures of *Panicum maximum* have been successfully cultured, with high plating efficiency, to obtain embryogenic callus, embryoids and plantlets (Lu *et al.*, 1981). In many instances the embryoids appeared to arise directly from embryogenic protoplasts.

Mesophyll protoplasts of cereals have never been successfully cultured. This, and the fact that even callus formation from the leaves of cereals has generally proven to be extremely difficult (Cocking, 1978), has created doubts about the totipotency of cereal leaf cells. Therefore, the recent success in inducing somatic embryogenesis leading to plant formation from cultured leaf segments of *Panicum maximum* (Lu and Vasil, 1981b), *Pennisetum purpureum* (Haydu and Vasil, 1981), and *Sorghum bicolor* (Wernicke and Brettell, 1980), is of considerable interest and significance, and clearly establishes the totipotency of the epidermal and mesophyll cells of cereal leaves. This also indicates that the difficulties currently being faced in the culture of cereal mesophyll protoplasts are technical in nature and are likely to be resolved.

While we have concentrated on obtaining protoplasts from morphogenetically competent cell cultures, others have depended largely on the manipulation of nutrient media, conditions for growth of donor plants, and methods of protoplast isolation and culture (Potrykus *et al.*, 1976; Galston, 1978; Potrykus, 1980). Several hundred thousand variations of the above conditions have been used. Unfortunately, none led to the growth of protoplasts leading to the regeneration of embryoids, shoots or plantlets. In most instances no growth was obtained. These experiences have led to suggestions that the lack of sustained cell division in grass protoplasts may be due to a mitotic block (Potrykus *et al.*, 1976; Kaur-Sawhney *et al.*, 1980). A gloomier forecast is made by Flores *et al.* (1981) who conclude that

"Considerable evidence leads to the belief that cereral protoplasts are constitutionally incapable of sustained division." On the basis of available evidence I believe these conclusions to be misleading, unfounded and unjustified. Our own success in the culture of *Panicum* and *Pennisetum* protoplasts, and the experiences of many others with nonmorphogenic cell cultures, shows unequivocally that not only do cereal/grass protoplasts undergo sustained cell divisions *in vitro*, but those isolated from morphogenetically competent cell cultures also form embryoids and plantlets.

A number of factors, some still undefined, are responsible not only for the initiation, but also for the maintenance, of embryogenic competence in callus as well as cell suspension cultures. A fuller understanding of these will be critical in establishing embryogenic suspension cultures from additional species for use in the isolation and culture of cereal protoplasts leading to plant regeneration.

In some of our recent studies with tissue cultures derived from immature embryos of *Triticum aestivum* we first have seen green "leafy" structures followed by the formation of a number of shoot meristems from the compact callus tissue present at the base of these structures (Ozias-Akins and Vasil, 1982). The shoot meristems later give rise to normal plants. Organized embryoids were formed only rarely in tissue cultures of wheat. The "leafy" structures have been interpreted to be the scutellum of precociously germinating embryos, in tissue cultures of *Triticum aestivum* (Ozias-Akins and Vasil, 1982), *Pennisetum americanum* (V. Vasil and Vasil, 1982a) and *P. purpureum* (Wang and Vasil, 1982). Somatic embryogenesis is initiated in all such cultures but organized embryoids are often not formed because of their precocious proliferation at an early stage of development, long before they can be morphologically recognized as embryos. If this indeed is what happens in tissue cultures of many Gramineae where leafy structures appear first (Tamura, 1968; Nakano and Maeda, 1979), then it may be possible to experimentally control and continue normal embryogenic development in other cereals and grasses also.

An important advantage of recovering plants
from tissue cultures through somatic embryogenesis
is that the somatic embryos, like their zygotic
counterparts, also arise from single cells
(Haccius, 1978). The embryos arise either from
single epidermal cells as in *Ranunculus sceleratus*
(Konar *et al.*, 1972), or from single cells present
in proembryonal masses in callus or cell suspension
cultures (Steward *et al.*, 1958, 1964; Halperin and
Jensen, 1967; Halperin, 1969; Backs-Husemann and
Reinert, 1970; Thomas *et al.*, 1972; McWilliam *et
al.*, 1974; Street and Withers, 1974; Tisserat and
DeMason, 1980). Similar evidence is now being
accumulated for the somatic embryos of cereals and
grasses (Lu and Vasil, 1981a; V. Vasil and Vasil,
1981b, 1982b).

The regeneration of plants by the process of
somatic embryogenesis also offers a number of other
advantages. In comparison to the two other major
methods of plant regeneration *in vitro* - the
formation of adventitious or axillary shoot buds -
many more plants can be obtained by somatic
embryogenesis. Indeed, each embryogenic cell is
potentially like a zygote and, under appropriate
conditions, may be induced to give rise to an
embryo (embryogenic cell suspension cultures
contain as many as $15-20 \times 10^6$ cells/ml). Plants
can be produced rapidly and in relatively short
periods of 3-6 weeks. Because such plants already
possess an integrated shoot-root system, they are
easy to establish in soil and grow rapidly into
adult plants. It is also possible to synchronize
the development of somatic embryos *in vitro*, either
by the physical assortment of specific developmental stages, or by the manipulation of culture
conditions. Such synchronized development is very
desirable for the establishment of automated
systems and mass propagation. Embryogenic cell
lines maintain their competence for long periods of
time, and give rise to genetically uniform and
normal plant populations. Embryogenic cell lines
can also be particularly useful in studying the
control of embryogenesis, in providing a suitable
source of embryogenic protoplasts for culture and
regeneration, and in developing single cell systems
for genetic modification experiments.

III. ANDROGENETIC HAPLOIDS

Haploid plants can be very useful in obtaining
homozygous and pure lines, and in the identification and expression of recessive genetic
characters. Such information can greatly assist in
mutation and breeding programs. Haploids can now
be obtained in many species – including most of the
important cereal species – by anther or isolated
microspore culture (I. K. Vasil, 1980c). Plant
regeneration takes place either through somatic
embryogenesis (rigorous histological evidence for
this is generally not available) or by the
organization of shoot meristems in callus
tissues. However, the incidence of haploid plant
formation in cereals is rather low and generally
less than 1% of the cultured anthers are
responsive. Many of the regenerated plants turn
out to be albinos (Sun *et al.*, 1979). Resolution
of these two problems is critical and urgent.
However, attempts can now be made to isolate
auxotrophic and other mutant cell lines from
haploid tissue cultures or plants of cereals.
Although some use of the anther-derived haploid
plants of cereals has been made in breeding
programs, particularly for rice in The People's
Republic of China and in Japan, further close
collaboration between plant breeders and those
engaged in the production of androgenetic haploids
is not only desirable, but essential.

IV. MUTANT/VARIANT AND RESISTANT CELL LINES AND
     PLANTS

Plant tissue cultures have been utilized to
isolate cell lines that are resistant to specific
drugs, antimetabolites, or toxins produced by
pathogenic organisms (Maliga, 1980). The
usefulness of such cell lines in the selection and
preferential growth of only the somatic hybrid
cells following protoplast fusion has already been
demonstrated (White and Vasil, 1979; Harms *et al.*,
1981).
There is, unfortunately, very little work of
this nature with cereal species. Cell cultures of
rice resistant to S-2-aminoethylcysteine were
isolated by Chaleff and Carlson (1975), and were

shown to have elevated levels of several amino
acids. Plants were not regenerated from this cell
line. Perhaps the most encouraging report on this
subject is that by Hibberd *et al*. (1980), who
isolated a cell line of maize that was resistant to
growth-inhibitory levels of lysine + threonine
(LT), and showed 2 to 9 fold increase in the pools
of lysine, threonine, isoleucine and methionine.
Maize plants regenerated from a stable LT cell line
were shown to retain resistance to LT, which was
shown to be inherited in the manner of a single
dominant nuclear gene (Hibberd and Green, 1982).
Analysis of free amino acids in kernels from LT
resistant plants showed 75-100 times increase in
the amount of free threonine, and about 4 times
increase in free serine, proline and methionine.
These results are signficant in that they show that
LT resistance selected *in vitro* was heritable, and
that it significantly affected amino acid
metabolism in tissue cultures, seedlings, and seeds
obtained from LT resistant plants.

Soils in many of the arid and coastal regions
of the world are unsuitable for agriculture owing
to their high salinity. One of the possible
remedies to this problem is the selection of salt
tolerant cell lines in tissue culture and
regeneration of plants from them. Although salt
tolerant cell lines have been selected *in vitro* it
is by no means clear whether the tolerance selected
in culture will be expressed at the whole plant
level. Recently tobacco plants regenerated from
salt tolerant cell lines selected *in vitro* have
been carried through three generations and have
been reported to retain their characteristics
(Nabors *et al.*, 1980).

Tissue culture selection has also been used to
obtain disease-resistant clones of sugarcane that
are resistant to *Helminthosporium sacchari*, Fiji
disease, and *Celerospora sacchari* (Heinz *et al.*,
1977). Similarly, cell lines and plants of maize
that are resistant to the toxin produced by
Helminthosporium maydis have been produced
(Gengenbach *et al.*, 1977; Brettell and Ingram,
1979; Brettell *et al.*, 1979, 1980a).

V.   SOMATIC HYBRIDIZATION

Interspecific as well as intergeneric
hybrids, which can not be obtained by conventional
breeding methods, have been produced by protoplast
fusion (Melchers *et al.*, 1978; Dudits *et al.*, 1980;
Gleba and Hoffmann, 1980; Schieder and Vasil,
1980).   However, somatic hybridization of cereal
species can not be currently accomplished owing to
the difficulties encountered in the regeneration of
plants from their protoplasts.

Work with model systems like *Datura* and
Nicotiana indicates that somatic hybrid cells
exhibit "hybrid vigour" (Schieder and Vasil, 1980),
and that traits for morphogenetic capacity are
dominant (Maliga *et al.*, 1977).   It will be
interesting, therefore, to see if protoplasts
obtained from a cell line that divides readily *in
vitro*, as in *Zea mays* (Potrykus *et al.*, 1979), and
fused with protoplasts which exhibit embryogenic
potential, as in *Pennisetum americanum* (V. Vasil
and Vasil, 1980) and *Panicum maximum* (Lu *et al.*,
1981), will divide and show morphogenic/embryogenic
potential.   Such an approach is also suggested by
division obtained in the fusion products of cell
suspension culture protoplasts of *Sorghum* that
divide *in vitro* and mesophyll protoplasts of maize
that do not divide in culture (Brar *et al.*, 1980).

One of the major limitations to the success of
somatic hybridization experiments is the lack of
discriminating selection systems for the recovery
of the few somatic hybrid cells formed as a result
of protoplast fusion (I. K. Vasil *et al.*, 1979a,
b).   Probably the most useful systems will be those
where selection is based on complementation between
cell lines resistant to antimetabolites, drugs,
etc., auxotrophic cell lines, as well as the
exploitation of "hybrid vigour" expressed by
somatic hybrids *in vitro*.

Protoplast fusion has been used to transfer
cytoplasmically based genetic traits, such as
cytoplasmic male sterility (Zelcher *et al.*, 1978;
Izhar and Power, 1979).   Recent evidence from
eukaryotic and prokaryotic organisms indicates that
susceptibility or resistance to at least some of
the drugs and pathotoxins may be controlled by
cytoplasmic genes.   Therefore, the production of
hybrids, which contain the mixture of two

cytoplasms but only one nuclear genome, can be
helpful not only in the transfer of cytoplasmic
genetic information from one plant to another, but
also in conferring drug or disease resistance and
in the understanding of nucleocytoplasmic
interactions.

   Helminthosporium maydis is an important
pathogen of *Zea mays*, and produces the toxin HmT.
It has been suggested that the mitochondria in the
susceptible plants are the primary site of HmT
action (York *et al.*, 1980). Earle and Gracen
(1980) have shown that resistance to HmT can be
conferred to the "hybrid" protoplasts by fusing
susceptible corn mesophyll protoplasts with soybean
protoplasts that are unaffected by the toxin.
Toxin resistance appears to act as a "dominant"
characteristic in hybrid cytoplasms. The
applicatin of this finding to cereals is currently
limited owing to the problems faced in the culture
of their protoplasts.

VI.   TRANSFORMATION

   Evidence for the successful and controlled
transformation of higher plant cells is still
lacking, in spite of the many claims to the
contrary which have been made during the past
decade and continue to be made today (Vasil *et al.*,
1979a; Kado and Kleinhofs, 1980). Considerable
interest, therefore, has been attracted by the
dramatic recent breakthroughs in the understanding
of the natural system of transformation and DNA
exchange provided by *Agrobacterium tumefaciens*,
which causes the neoplastic crown gall disease in
many dicotyledonous plants, because of the
possibility that it may serve as a natural vector
for the introduction of new genetic information
into plant cells (Nester *et al.*, 1977; Schell and
Von Montagu, 1977; Schell *et al.*, 1979; Braun,
1980; Yang *et al.*, 1980). The most recent
investigations suggest that DNA from the Ti plasmid
is present in the nuclear fraction of crown gall
tumor cells (Chilton *et al.*, 1980; Willmitzer *et
al.*, 1980). *In vitro* transformation of cultured
tobacco cells with *A. tumefaciens* (Marton *et al.*,
1979), and of petunia protoplasts by isolated Ti
plasmid (Davey *et al.*, 1980), has been reported,

and this further testifies to the potential of the
Ti plasmid serving as a vector for the transformation of plant cells. Crown gall tissues of
tobacco have been induced to form plants (Braun and
Wood, 1976; Sacristan and Melchers, 1977).

Agrobacterium tumefaciens does not infect
monocotyledonous plants, including cereal species,
and it remains to be seen whether the Ti plasmid
system can be used as a vehicle for the transformation of cereals. Tumor induction by
Agrobacterium involves the attachment of the
bacterium to a site on the host plant cell wall
(Lippincott et al., 1977; Matthysse et al., 1978;
Whatley et al., 1978). The absence of specific
host sites on the cell walls of monocotyledonous
species might, therefore, be responsible for the
inability of Agrobacterium to infect monocotyledonous species. It would be important to
test this hypothesis by trying to infect
protoplasts of monocots, but particularly cereals,
by Agrobacterium or with T-DNA. Success in such
efforts will make it possible to attempt
transformation of cereals. The use of liposomes
(Paphadjopoulos et al., 1980) as vehicles may
confer further advantage because it will provide a
better and more efficient method for the transfer
of the T-DNA in an intact form to the plant cell.

VII. MOLECULAR BIOLOGY

A fuller and detailed discussion of the current
status and future prospects of the molecular
biology of cereals is available from other
sources. My comments, therefore, are restricted to
those areas where an interaction between the
techniques of plant cell culture and molecular
biology is now feasible and desirable.

Cereal crops have provided one of the most
important sources of human nutrition from the
beginnings of human civilization, and will continue
to do so in the future. It is not surprising,
therefore, that they have been the favorite object
of study by plant breeders and geneticists.
Conventional methods of cytogenetics and breeding
have provided very useful and valuable information
about the genetics of cereals, including location
of many genes at specific sites on the
chromosomes. Although such work has been done with

many cereal species, the bulk of the available
information is about maize (Neuffer *et al.*,
1968). It is only recently that a variety of
molecular techniques, collectively and popularly
known as recombinant DNA techniques, have been used
to characterize, clone, and study the organization
and expression of nuclear as well as organellar
genomes of cereals (Bedbrook *et al.*, 1980; Bogorad
et al., Burr and Burr, 1980; Flavell, 1980; Flavell
et al.,1980; Hake and Walbot, 1980; Leaver, 1980;
Leaver and Forde, 1980; Rimpau *et al.*, 1980).

Studies involving cell culture, somatic cell
genetics and molecular biology are beginning to
provide an interesting and deeper understanding of
certain nucleo-cytoplasmic interactions in the
control of gene expression and plant development.
Some of the reported variability observed in
cytoplasmic traits may be caused by the
recombination or loss of mtDNA, mutations, or
intracellular heterogeneity of mtDNA (Gengenbach *et
al.*, 1977, 1980; Belliard *et al.*, 1979; Brettell *et
al.*, 1979, 1980a; Levings *et al.*, 1980).

Many unanswered questions remain. Nuclear
restorer genes, which restore fertility in
cytoplasmic male sterile plants, have been known
for a long time. Episome-like fertility elements
have been postulated in maize (Laughnan and Gabay,
1975). Where are these located, and how do these
function? Can they move from the cytoplasm and be
inserted into chromosomal DNA or *vice versa*? The
molecular role and interaction of the nuclear gene
products in regulating the expression of maize
mitochondrial genome in cytoplasmic male sterility,
and in the restoration of fertility, is not
understood. The existence of plasmid-like DNA's in
the mitochondria of maize has been demonstrated,
and the possible involvement of at least one of
these in reversion to male fertility in cms-S maize
plants is indicated (Kemble and Bedbrook, 1980;
Koncz *et al.*, 1980; Levings *et al.*, 1980).

The transposable controlling elements of maize
have been shown to cause mutations and control the
activity and expression of genes (McClintock, 1956;
Fincham and Sastry, 1974). Controlling elements
have also been described and studied in a number of
eukaryotes and prokaryotes. They are thought to
consist of transposable DNA sequences. The
components of one of the controlling element

systems in maize have been shown to be active in
callus tissue cultures derived from the endosperm
(Gorman and Peterson, 1978). This may further help
in the manipulation of the activity of the
controlling elements, the isolation of their
products, and in the elucidation of the molecular
basis of controlling element action.

The controlling elements, the fertility
elements, and the episomal plasmid-like DNA's found
in the mitochondria of maize, probably all function
as transposable elements by integrating into mtDNA
and/or nuclear DNA. If such integration does
indeed take place, can it be controlled and
experimentally manipulated to the extent that the
controlling/fertility/plasmid-like elements can be
isolated, characterized, genetically modified by
the insertion of foreign DNA, cloned, and then used
as vectors for the transformation of plant,
particularly cereal, cells?

The preceding pages provide strong and
convincing evidence of the rapid, encouraging and
important progress in cell culture, somatic cell
genetics and molecular biology of cereal plants.
It is hoped that this will provoke some productive
discussion and will help attract other scientists
to this challenging and potentially useful area of
research. I would also like to use this
opportunity to make a plea for greater cooperation
and interaction between those of use interested in
plant cell and tissue culture, with agronomists,
plant breeders and molecular biologists and
geneticists, because that is essential and critical
for the rapid achievement of many of our shared
goals.

ACKNOWLEDGMENTS

My discussions with a number of colleagues and
friends, particularly during an extensive lecture
tour of Europe in the summer of 1980, and Australia
and the The Peoples Republic of China in 1981, have
greatly helped me in the preparation of this
review. Especially rewarding and challenging have
been the frequent spirited discussions with my co-
workers, Peggy Ozias-Akins, Claudia Botti, Charlene
Boyes, Zsolt Haydu, Wai-Jane Ho, Chin-yi Lu,
Valerie Pence, T. S. Rangan, Vimla Vasil, Da-Yuan

Wang, Derek White and Martin Wilson. I thank the
following for helping to make this report more
timely and up to date by generously providing
preprints and other important information: Dr.
Richard B. Flavell (Plant Breeding Institute,
Cambridge, England), the late Dr. Emrys Thomas
(Rothamsted Experiment Station, Harpenden,
England), Dr. Richard I. S. Brettell, Dr. Patrick
J. King, Dr. Ingo Potrykus and Dr. Wolfgang
Wernicke (Freidrich-Miescher-Institut, Basel,
Switzerland), Dr. C. John Jensen (Riso National
Laboratory, Roskilde, Denmark), Dr. Glen B. Collins
(University of Kentucky, Lexington, KY), Dr.
Elizabeth D. Earle (Cornell University, Ithaca,
NY), Dr. Charles E. Green (University of Minnesota,
Minneapolis, MN), Dr. James F. Shepard (Kansas
State University, Manhattan, KS), and Dr. Larkin C.
Hannah and Dr. Rusty J. Mans (University of
Florida, Gainesville, FL).

REFERENCES

Backs-Husemann, D., and Reinert, J. (1970).   *Proto-
    plasma 70*, 49-60.
Bedbrook, J., Gerlach, W., Smith, S., Jones, J.,
    and Flavell, R. (1980).   *In* "Genome
    Organization and Expression in Plants" (C. J.
    Leaver, ed.).   Plenum Press, New York.
Belliard, G., Vedel, F., and Pelletier, G.
    (1979).   *Nature 281*, 401-403.
Bennici, A., and D'Amato, F.   (1978).   *Z.
    Pflanzenzuchtg. 81*, 305-311.
Bogorad, L., Jolly, S. O., Kidd, G., Link, G., and
    McIntosh, L.   (1980).   *In* "Genome Organization
    and Expression in Plants," (C. J. Leaver,
    ed.).   Plenum Press, New York.
Brar, D. S., Rambold, S., Constabel, F., and
    Gamborg, O. L. (1980).   *Z. Pflanzenphysiol. 96*,
    269-275.
Braun, A. C. (1980). *In Vitro 16*, 38-48.
Braun, A. C., and Wood, H. N.   (1976).   *Proc. Nat.
    Acad. Sci. U.S.A. 73*, 496-500.
Brettell, R. I. S., Goddard, B. V. D., and Ingram,
    D. S. (1979).   *Maydica 24*, 203-213.
Brettell, R. I. S., and Ingram, D. S. (1979).
    Biol. Rev. 54, 329-345.

Brettell, R. I. S., Thomas, E., and Ingram, D. S. (1980a). *Theoret. Appl. Genet.* 58, 55-58.
Brettell, R. I. S., Wernicke, W., and Thomas E. (1980b). *Protoplasma 104*, 191-198.
Burr, F. A., and Burr, B. (1980). *In* "Genome Organization and Expression in Plants," (C. J. Leaver, ed.). Plenum Press, New York.
Cai, Q., Quian, Y., Zhou, Y., and Wu, S. (1978). *Acta Bot. Sinica 20*, 97-102.
Chaleff, R. S., and Carlson, P. S. (1975). *In* "Modification of the Information Content of Plant Cells" (R. Markham, D. R. Davies, D. A. Hopwood, and R. W. Horne, eds.). American Elsevier, New York.
Chilton, M. D., Saiki, R. K., Yadav, N., Gordon, M. P., and Quetier, F. (1980). *Proc. Nat. Acad. Sci. U.S.A. 77*, 4060-4064.
Cocking, E. C. (1978). *In* "Proceedings of Symposium on Plant Tissue Culture." Science Press, Peking.
Coe, E. H., Jr., and Neuffer, M. G. (1978). *In* "The Clonal Basis of Development" (S. Subtelny and I. M. Sussex, eds.). Academic Press, New York.
Crooks, D. M. (1933). *Bot. Gaz. 95*, 209-239.
Dale, P. J. (1980). *Z. Pflanzenphysiol. 100*, 73-77.
Dale, P. J., Thomas, E., Brettell, R. I. S., and Wernicke, W. (1981). *Pl. Cell. Tiss. Org. Cult. 1*, 47-55.
Davey, M. R., Cocking, E. C., Freeman, J., Pearce, N., and Tudor, I. (1980). *Pl. Sci. Let. 18*, 307-313.
Deka, P. C., and Sen, S. K. (1976). *Molec. Gen. Genet. 145*, 239-243.
Dudits, D., Fejer, O., Hadlaczky, G., Koncz, C., Lazar, G. B., and Horvath, G. (1980). *Molec. Gen. Genet. 145*, 283-288.
Dunstan, D. I., Short, K. C., Dhaliwal, H., and Thomas, E. (1979). *Protoplasma 101*, 355-361.
Earle, E. D., and Gracen, V. E. (1979). *Pl. Physiol. Suppl.* 136.
Fincham, J. R. S., and Sastry, G. R. K. (1974). *Ann. Rev. Genet. 8*, 15-50.
Flavell, R. (1980). *Ann. Rev. Pl. Physiol. 31*, 569-596.
Flavell, R., Rimpau, J., Smith, D. N., O'Dell, M., and Bedbrook, J. R. (1980). *In* "Genome

Organization and Expression in Plants" (C. J.
Leaver, ed.). Plenum Press, New York.
Flores, H. E., Kaur-Sawhney, R., and Galston, A.
W. (1981). *Adv. Cell Cult. 1*, 241-279.
Galston, A. W. (1978). *In* "Propagation of Higher
Plants Through Tissue Culture" (K. W. Hughes,
R. Henke, M. Constantin, eds.). U.S. Dept. of
Energy, Oak Ridge, TN.
Gengenbach, B. G., Connelly, J. A., Pring, D. R.,
and Conde, M. F. (1980). *Theoret. Appl.
Genet. 59*, 161-167.
Gengenbach, B. G., Green, C. E., and Donovan, C.
M. (1977). *Proc. Nat. Acad. Sci. U.S.A. 74*,
5113-5117.
Gleba, Y. Y., and Hoffmann, F. (1980). *Planta
149*, 112-117.
Gorman, M. B., and Peterson, P. A. (1978).
Maydica 23, 173-186.
Green, C. E., and Phillips, R. L. (1975). *Crop
Sci. 15*, 417-421.
Gresshoff, P. M. (1980). *Bot. Gaz. 141*, 157-164.
Haccius, B. (1978). *Phytomorphology 28*, 74-81.
Hake, S., and Walbot, V. (1980). *Chromosoma 79*,
251-270.
Halperin, W. (1969). *Ann. Rev. Pl. Physiol. 20*,
395-418.
Halperin, W., and Jensen, W. A. (1967). *J.
Ultrastruc. Res. 18*, 426-443.
Harms, C. T., Potrykus, I., and Widholm, J. M.
(1981). *Z. Pflanzenphysiol. 101*, 377-390.
Haydu, Z., and Vasil, I. K. (1981). *Theoret.
Appl. Genet. 59*, 269-273.
Heinz, D. J., Krishnamurthy, M., Nickell, L. G.,
and Maretzki, A. (1977). *In* "Plant Cell,
Tissue, and Organ Culture" (J. Reinert and Y.
P. S. Bajaj, eds.). Springer-Verlag, New York.
Hibberd, K. A., and Green, C. E. (1982). *Proc.
Nat. Acad. Sci. U.S.A.* (In Press).
Hibberd, K. A., Walker, T., Green, C. E., and
Gengenbach, B. G. (1980). *Planta 148*, 183-
187.
Izhar, S., and Power, J. B. (1979). *Pl. Sci.
Let. 14*, 49-55.
Kado, C. I., and Kleinhofs, A. (1980). *In*
"Perspectives in Plant Cell and Tissue Culture"
(I. K. Vasil, ed.) Int. Rev. Cytol. Suppl.
11B. Academic Press, New York.

Kao, K. N., and Michayluk, M. R. (1980). *Z. Pflanzenphysiol. 96*, 135–141.

Kaur-Sawhney, R., Flores, H. E., and Galston, A. W. (1980). *Pl. Physiol. 65*, 368–371.

Kemble, R. J., and Bedbrook, J. R. (1980). *Nature 284*, 565–566.

Koblitz, H. (1976). *Biochem. Physiol. Pfl. 170*, 287–293.

Konar, R. N., Thomas, E., and Street, H. E. (1972). *J. Cell Sci. 11*, 77–93.

Koncz, C., Sumegi, J., Sain, B., Kalman, L., and Dudits, D. (1980). *In* "DNA-Recombination, Interactions and Repair" (S. Zadrizil and J. Sponar, eds.) Pergamon Press, New York.

Laughnan, J. R., and Gabay, S. J. (1975). *In* "Genetics and Biogenesis of Cell Organelles" (C. W. Birky, P. S. Perlman and T. J. Byers, eds.). Ohio State University Press, Columbus.

Leaver, C. J. (ed.). (1980). "Genome Organization and Expression in Plants." Plenum Press, New York.

Leaver, C. J., and Forde, B. G. (1980). *In* "Genome Organization and Expression in Plants" (C. J. Leaver, ed.). Plenum Press, New York.

Levings, C. S., Kim, B. D., Pring, D. R., Conde, M. F., Mans, R. J., Laughnan, J. R., and Gabay-Laughnan, S. J. (1980). *Science 209*, 1021–1023.

Lippincott, B. B., Whatley, M. H., and Lipincott, J. A. (1977). *Pl. Physiol. 59*, 388–390.

Lu, C., and Vasil, I. K. (1981a). *Ann. Bot. 48*, 543–548.

Lu, C., and Vasil, I. K. (1981b). *Theoret. Appl. Genet. 59*, 275–280.

Lu, C., and Vasil, I. K. (1982). *Amer. J. Bot. 69*, 77–81.

Lu, C., Vasil, V., and Vasil, I. K. (1981). *Z. Pflanzenphysiol. 104*, 311–318.

Lu, C., Vasil, I. K., and Ozias-Akins, P. (1982). *Theoret. Appl. Genet.* In Press.

Maliga, P. (1980). *In* "Perspectives in Plant Cell and Tissue Culture" (I. K. Vasil, ed.). Int. Rev. Cytol. Suppl. 11A. Academic Press, New York.

Maliga, P., Lazar, G., Joo, F., Nagy, A. H., and Menczel, L. (1977). *Molec. Gen. Genet. 157*, 291–296.

Marton, L., Wullems, G. J., Molendijk, L. and
    Schilperoort, R. A.   (1979). *Nature 277*, 129-
    131.
Matthysse, A., Wyman, P., and Holmes, K.
    (1978). *Infect. Immun. 22*, 516-520.
McClintock, B. (1956). *Cold Spring Harbor Symp.
    Quant. Biol. 21*, 197-216.
McCoy, T. J., and Bingham, E. T. (1977). *Pl.
    Sci. Let. 10*, 59-66.
McWilliam, A. A., Smith, S. M., and Street, H.
    E. (1974). *Ann. Bot. 38*, 243-250.
Melchers, G., Sacristan, M. D., and Holder, A.
    A. (1978). *Carlsberg Res. Comm. 43*, 203-218.
Mix, G., Wilson, H. M., and Foroughi-Wehr, B.
    (1978). *Z. Pflanzenzuchtg. 80*, 89-99.
Nabors, M. W., Gibbs, S., Bernstein, C., and
    Meiss, M. (1980). *Z. Pflanzenphysiol. 97*, 13-
    17.
Nakano, H., and Maeda, E. (1979). *Z.
    Pflanzenphysiol. 93*, 449-458.
Nemet, G., and Dudits, D. (1977). *In* "Use of
    Tissue Cultures in Plant Breeding" (F. J.
    Novak, ed.). Czechoslovak Academy of Sciences,
    Institute of Experimental Botany, Prague.
Nester, E. W., Merlo, D. J., Drummond, M. H.,
    Sciaky, D., Montoya, A. L., and Chilton, M.
    D. (1977). *In* "Genetic Engineering for
    Nitrogen Fixation" (A. Hollaender, R. H.
    Burris, P. R. Day, R. W. F. Hardy, D. R.
    Helinski, M. R. Lamborg, L. Owens and R. C.
    Valentine, eds.). Plenum Press, New York.
Neuffer, M. G., Jones, L., and Zuber, M. S.
    (1968). "The Mutants of Maize." Crop Sci. Soc.
    America, Madison, WI.
Ogura, H. (1976). *Jap. J. Genet. 51*, 161-174.
Ozias-Akins, P. and Vasil, I. K. (1982).
    Protoplasma
Paphadjopoulos, D., Wilson, T., and Taber, R.
    (1980). *In Vitro 16*, 49-54.
Phillips, G. C., and Collins, G. B. (1980). *Crop
    Sci. 20*, 323-326.
Potrykus, I. (1980). *In* "Advances in Protoplast
    Research" (L. Ferenczy, G. L. Farkas and G.
    Lazar, eds.). Akademiai Kiado, Budapest.
Potrykus, I., Harms, C. T., and Lorz, H.
    (1976). *In* "Cell Genetics in Higher Plants"
    (D. Dudits, G. L. Farkas, and P. Maliga,
    eds.). Akademiai Kiado, Budapest.

Potrykus, I., Harms, C. T., and Lorz, H.
(1979). *Theoret. Appl. Genet. 54*, 209-214.
Rangan, T. S. (1974). *Z. Pflanzenphysiol. 72*,
456-459.
Rimpau, J., Smith, D. B., and Flavell, R. B.
(1980). *Heredity 44*, 131-149.
Sacristan, M. D., and Melchers, G. (1977).
Molec. Gen. Genet. 152, 111-117.
Saunders, J. W., and Bingham, E. T. (1972). *Crop
Sci. 12*, 804-809.
Schell, J., and Van Montagu, M. (1977). *In*
"Genetic Engineering for Nitrogen Fixation" (A.
Hollaender, R. H. Burris, P. R. Day, R. W. F.
Hardy, D. R. Helinski, M. R. Lamborg, L. Owens
and R. C. Valentine, eds.). Plenum Press, New
York.
Schell, J., Van Montagu, M., De Beuckeleer, M.,
De Block, M., Depicker, A., De Wilde, M.,
Engler, G., Genetello, C., Hernalsteens, J. P.,
Holsters, M., Seurinck, J., Silva B., Van
Vliet, F., and Villarroel, R. (1979). *Proc.
R. Soc. London, B 204*, 251-266.
Schieder, O., and Vasil, I. K. (1980). *In*
"Perspectives in Plant Cell and Tissue Culture"
(I. K. Vasil, ed.). Int. Rev. Cytol. 11B.
Academic Press, New York.
Shepard, J. F., Bidney, D., and Shahin, E.
(1980). *Science 208*, 17-24.
Shimada, T., and Yamada, Y. (1979). *Japan J.
Genet. 54*, 379-385.
Springer, W. D., Green, C. F., and Kohn, K. A.
(1979). *Protoplasma 101*, 269-281.
Sree Ramulu, K., Devreux, M., Ancora, G., and
Laneri, U. (1976). *Z. Pflanzenzuchtg. 76*,
299-319.
Steffenson, D. M. (1968). *Amer. J. Bot. 55*, 354-
369.
Steward F. C., Mapes, M. O., and Mears, K.
(1958). *Amer. J. Bot. 45*, 705-708.
Steward, F. C., Mapes, M. O., Kent, A. E., and
Holsten, R. D. (1964). *Science 143*, 20-27.
Street, H. E., and Withers, L. A. (1974). *In*
"Tissue Culture and Plant Science 1974" (H. E.
Street, ed.). Academic Press, New York.
Sun, C. S., Wu, S. C., Wang, C. C., and Chu, C.
C. (1979). *Theoret. Appl. Genet. 55*, 193-197.
Tamura, S. (1968). *Proc. Japan Acad. 44*, 544-548.

Thomas, E., King, P. J., and Potrykus, I.
    (1977). *Naturwiss. 64,* 587.
Thomas, E., King, P. J., and Potrykus, I.
    (1979). *Z. Pflanzenzuchtg. 82,* 1-30.
Thomas, E., Konar, R. N., and Street, H. E.
    (1972). *J. Cell. Sci. 11,* 95-109.
Tisserat, B., and De Mason, D. A. (1980). *Ann.*
    Bot. 46, 465-472.
Vasil, I. K. (ed.). 1980a. "Perspectives in
    Plant Cell and Tissue Culture." Int. Rev.
    Cytol., Suppl. 11A. Academic Press, New York.
Vasil, I. K. (ed.). 1980b. "Perspectives in
    Plant Cell and Tissue Culture" Int. Rev.
    Cytol., Suppl. 11B. Academic Press, New York.
Vasil, I. K. 1980c. *In* "Perspectives in Plant
    Cell and Tissue Culture." (I. K. Vasil,
    ed.). Int. Rev. Cytol. Suppl. 11A. Academic
    Press, New York.
Vasil, I. K., Ahuja, M. R., and Vasil, V.
    (1979a). *Adv. Genet. 20,* 127-215.
Vasil, I. K., and Hildebrandt, A. C. (1966).
    Amer. J. Bot. 53, 860-869.
Vasil, I. K., and Vasil V. (1980a). *In*
    "Perspectives in Plant Cell and Tissue Culture"
    (I. K. Vasil, ed.). Int. Rev. Cytol. Suppl.
    11A, Academic Press, New York.
Vasil, I. K., and Vasil V. (1980b). *In*
    "Perspectives in Plant Cell and Tissue
    Culture." (I. K. Vasil, ed.). Int. Rev.
    Cytol., Suppl. 11B. Academic Press, New York.
Vasil, I. K., Vasil V., White, D. W. R., and Berg,
    R. H. (1979b). *In* "Plant Regulation and
    World Agriculture". (T. K. Scott, ed.).
    Plenum Press, New York.
Vasil, V., Vasil, I. K. (1979). *Z.*
    Pflanzenphysiol. 92, 379-383.
Vasil, V., Vasil, I. K. (1980). *Theoret. Appl.*
    Genet. 56, 97-99.
Vasil, V., Vasil, I. K. (1981a). *Amer. J. Bot.*
    68, 864-872.
Vasil, V., Vasil, I. K. (1981b) *Ann. Bot. 47,*
    669-678.
Vasil, V., Vasil, I. K. (1982a). *Amer. J. Bot.*
    69,
Vasil, V., Vasil, I. K. (1982b).
Wang, D., and Vasil, I. K. (1982). *Pl. Sci. Let.*
Wernicke, W., and Brettell, R. (1980). *Nature*
    287, 138-139.

Whatley, M H., Margot, J. B., Schell, J.,
    Lippincott, B. B., and Lippincott, J. A.
    (1978). *J. Gen. Microbiol. 107*, 395.
White, D. W. R., and Vasil, I. K. (1979).
    Theoret. Appl. Genet. 55, 107–112.
Willmitzer, L., De Benckleleer, M., Lemmers, M.,
    Van Montagu, M., and Schell, J. (1980).
    Nature 287, 359–360.
Yang, F., Montoya, A. L., Nester, W., and Gordon,
    M. P. (1980). *In Vitro 16*, 87–92.
York, D. W., Earle, E. D., and Gracen, V. E.
    (1980). *Canad. J. Bot. 58*, 1562–1570.
Zelcher, A., Aviv, D., and Galun, E. (1978). *Z.
    Pflanzenphysiol. 90*, 397–407.

CHAPTER 10

SOMATIC CELL FUSION FOR INDUCING
CYTOPLASMIC EXCHANGE: A NEW BIOLOGICAL
SYSTEM FOR CYTOPLASMIC GENETICS IN HIGHER PLANTS

Esra Galun[1]

Department of Plant Genetics
The Weizmann Institute of Science,
Rehovot, Israel

In the following short review I shall first survey
studies dealing with somatic hybridization which furnished
information on chloroplast and mitochondrial assortment and
transmission. Some of these studies will then be analysed
in order to clarify the approaches of the various research
groups. I shall also focus on several experiments with the
"donor-recipient" protoplast fusion system developed in our
laboratory, and conclude with perspectives of cell genetics
and applied research objectives which may be approached by
cytoplasmic hybridization in higher plants.

PLASTIDS AND CHLOROPLASTS IN SOMATIC HYBRIDS AND CYBRIDS

Bridging the gap between species which cannot be hybridized sexually was probably the primary aim of pioneering
investigators who attempted somatic hybridization in angiosperms (see references in Schieder and Vasil, 1980). Nevertheless, awareness of the importance of somatic hybridization
in respect to chloroplast transmission became evident with
the first success with this technique. Plants resulting from
fusion between protoplasts having different chloroplasts
(i.e. *Nicotiana langsdorffii* and *N. glauca*) were found to be
composed (with only one exception) of uniform chloroplast

---

[1] *Irene and David Schwartz Professor of Plant Genetics*

PLANT IMPROVEMENT
AND SOMATIC CELL GENETICS

205

populations of either one of the two fusion partners (Kung
et al., 1975; Chen et al., 1977). This indicated an abrupt
sorting-out from the initial heteroplastomic fusion product.
Although a complete sorting-out of chloroplasts in plants
resulting directly from protoplast fusion, or in their sexual
progeny, was commonly observed, there were notable exceptions.
Theoretical model building based on such sorting-out seems
premature as we lack information on relevant parameters,
such as number of chloroplasts (or plastids) per protoplast
in each of the fusion partners, mean number of genomes in
each chloroplast at the time of fusion, replication potential
of the chloroplast genomes in cultured fusion products, and
the number of cell divisions between initial fusion and
complete sorting-out. We should thus wait for more information on these parameters before attempting to base the sorting
out of chloroplasts on statistical, physiological or genetic
arguments. Obviously, somatic hybridization is not the only
means to combine a given nuclear genome with a specific
chloroplast population (whether for genetic research or
breeding purposes). As pointed out rightfully by Melchers
(1980), biparental chloroplast transmission was recorded in
several plant species (see Galun, 1981) as a potential source
for organelle assortment. Therefore heteroplastomic zygotes
should not be ruled out. But, transmission of chloroplasts
via sexual crosses involves also nuclear-gene recombinations,
rendering the establishment of plants, homozygous in respect
to both nuclear and plastid genomes, rather tedious (Galun
and Aviv, 1978).
    Table 1 lists 12 fusion systems in which the chloroplast
compositions of the resulting hybrids or cybrids were analyzed.
Two of the early reported systems (N. glauca + N. langsdorffii
and L. esculentum + S. tuberosum) were conducted to achieve
interspecific and intergeneric hybridization respectively,
thus the analyses of chloroplast compositions were supple-
mental endeavours. Later, chloroplast-controlled traits, e.g.,
plastome-controlled chloroplast defects (Gleba, 1979), and
plastome-controlled drug resistances (Menczel et al., 1981b)
were used as cytoplasmic markers. Once efficient techniques
to identify the chloroplast compositions became available,
studies aimed specifically at the utilization of protoplast
fusion for organelle transfer were initiated. Some of these
techniques, such as isoelectric focussing of the polypeptides
of ribulose-1,5-bisphosphate carboxylase (RUBPCase) (von
Wettstein et al., 1978) and chloroplast DNA restriction
patterns (see Fluhr and Edelman, 1981) are of general
applicability but require multistep biochemical procedures.
Others, as tentoxin or streptomycin resistance (see Galun,
1981), involve simple tests but are applicable only to

TABLE 1: Main Systems of Somatic Hybridization in Higher Plants I. Chloroplast Analysis

Fusion Partners	Means of analyses	Authors
1. *N. glauca*(glc) + *N. longsforffii* (lng)	IEF of RUBPCase	Kung et al. (1975); Chen et al. (1977)
2. *L. esculentum*(esc) + *S. tuberosum*(tbr)	IEF of RUBPCase	Melchers et al.(1978)
3. *N. sylvestris*(tbc) + *N. tabacum* -CMS (L-92)*	IEF of RUBPCase,tentoxin Chl DNA	Zelcer et al. (1978); Aviv et al. (1980)
4. *N. tabacum*(tbc) + *N. tab.* var. techne(dbu)	Chl DNA	Belliard et al.(1978)
5. *N. tabacum*(tbc-P⁻) + *N. tabacum*(tbc-P⁺)	Variegation,IEF of RUBPCase	Gleba (1979)
6. *N. knightiana* (kng) + *N. sylvestris*(tbc-kᴿ-Pᐨ)	Kanamycin, greening	Menczel et al.(1978)
7. *N. tabacum*(tbc) + *N. sylvestris*-CMS(L-92)*	Tentoxin, Chl DNA	Aviv et al. (1980)
8. *N. sylvestris*(tbc) + *N. tabacum*(tbc-SRᴿ)	Streptomycin	Medgyesy et al(1980)
9. *N. tabacum*(tbc) + *N. rustica*(rsᵗ)	IEF of RUBPCase	Iwai et al.(1980)
10. *N. debneyi*(dbnᵃ)+*N. debneyi*(dbnᵇ)	Chl DNA	Scowcroft and Larkin (1981)
11. *N. tabacum*(tbc-SRᴿ) + *N. knightiana*(kng)	Streptomycin, Chl DNA	Menczel et al.(1981a)
12. *N. tabacum*(tbc-SRᴿ) + *N.plumbaginifolia* (plm)*	Streptomycin,Chl DNA	Sidorov et al. (1981) Menczel et al.(1981b)

*
X-irradiated protoplasts

Abbreviations: abbreviations in parenthesis indicate chloroplast genomes, i.e. (glc)= *N. glauca* chloroplasts; CMS–cytoplasmic male sterility; Kᴿ–kanamycin resistance; L-92–chloroplast genome alien to *N.tabacum* (see Aviv et al. 1980 for details); P⁻–mutated (albino) chloroplasts; SRᴿ– streptomycin resistance; IEF–isoelectric focussing; RUBPCase–ribulose-1,5-bisphosphate carboxylase.

specific fusion partners. These identification techniques
were recently augmented with other methodological
improvements to facilitate chloroplast transfer by protoplast
fusion. Two of these will be mentioned here. We developed
the "donor-recipient" system in which the "donor" protoplasts
are X-irradiated before fusion with non-irradiated "recipient"
protoplasts (see below). Maliga and his associates successfully utilized iodoacetate treatment which renders protoplasts
incapable of division unless they are immediately fused with
non-treated protoplasts (Medgyesy et al., 1980). This
technique was subsequently combined with the "donor-recipient"
system (Sidorov et al.,) to further improve chloroplast
transfer among plant species.

MALE STERILITY AND MITOCHONDRIAL ASSORTMENT IN SOMATIC
HYBRIDS AND CYBRIDS

    The fate of mitochondria and mitochondrial genomes in
hybrids, cybrids and their respective progenies, following
fusion between protoplasts having different mitochondrial
populations, was studied in only four systems (Table 2).
Belliard et al. (1979) were the first to indicate that hybrid
plants, resulting from fusion between protoplasts having
different mitochondria, contain mitochondrial genomes which
according to their mt DNA restriction patterns were not
identical to either of the parental mt DNAs. The hybrids'
mtDNA restriction patterns were similar, but not identical
to one or the other of the restriction patterns of the parents
and also contained new DNA fragments. These authors
suggested that DNA recombination caused the novel restriction
patterns, though other explanations were not ruled out. A
similar observation was made recently by Nagy et al.(1981).
Mitochondrial genomes were traced also in two other fusion
systems studied in our laboratory. These systems shall be
detailed in a following section.
    Both Belliard et al.(1979) and Galun et al(1981) found
correlations between the mitochondrial genomes and cytoplasmic
male sterility in the cybrid progenies. A detailed study on
the transmission of male sterility by somatic hybridization
was conducted by Izhar and collaborators (Table 2). These
investigators were studying the plasmones of Petunia somatic
hybrids (Izhar and Power, 1979; Izhar and Tabib, 1980; S.
Izhar personal communication). Their system differed
apparently from the above mentioned Nicotiana systems. A
continuous heteroplasmonic condition is maintained not only in
the Petunia plants regenerated directly from the fusion

TABLE 2. Main Systems of Somatic Hybridization in Higher Plants II. Analysis of Mitochondria and Male Sterility

Fusion Parents	Analysis	Authors
N. tabacum + *N. tabacum* v. techne (CMS) (tbc/tbc + dbn/dbn)	Male sterility/fertility, mtDNA	Belliard et al. (1979)
Petunia axillaris + *P. hybrida* (CMS) (no information on chl and mt)	Male sterility/fertility	Izhar and Power (1979 Izhar and Tabib (1980)
N. sylvestris + *N. tabacum*(CMS) *	Male sterility/fertility	Aviv et al. (1980); Galun and Aviv (1979) Galun et al. (1981)
N. sylvestris(CMS) + *N.tabacum* * (L-92/Type-A + tbc/tbc)	Male sterility/fertility ** mtDNA	Aviv and Galun (1980); Galun et al.(1981)
N. tabacum + *N. knightiana* (tbc-SR[R]/tbc+kng/kng)	mtDNA	Nagy et al. (1981)

* X-irradiated protoplasts;

** Individual cybrids and their progenies were tested for both chloroplast and mitochondrial composition.

Abbreviations: abbreviations in parenthesis indicate chloroplast and mitochondrial genomes (i.e. dbn/dbn=N. debneyi chloroplasts and N.debneyi mitochondria); Type-A - see Aviv and Galun, 1980; other abbreviations as in Table 1.

product (cytoplasmic male sterile, CMS, *Petunia hybrida* +
fertile *P. axillaris*) but also through sexual reproduction,
with sorting-out occurring in some of the F_2 and F_3
generation plants. The control of CMS in *Petunia* was not yet
assigned to specific cytoplasmic elements. Hence, Izhar and
collaborators used, in their published work, the phenotypic
expression (male sterility or fertility) of selfed and test-
cross progenies, to characterize the plasmones. We should
therefore wait for information on the mitochondrial and
chloroplast compositions of the cybrid progenies to better
understand the fate of organelles in *Petunia* cybrids and their
progenies.

THE "DONOR-RECIPIENT" SYSTEM

In our early efforts to establish a cell-genetic system by
protoplast culture and fusion in *Nicotiana* we encountered the
following problem. No division of cultured protoplasts
occurred unless plating was at a certain range of cell density.
This lack of division capacity of sparsely plated protoplasts
was very disturbing when only a small protoplast number was
available or when due to certain treatments, only a small
number of viable protoplasts was retained. To overcome this
problem we developed the feeder-layer technique (Raveh *et al.*,
1973). In this technique protoplasts are exposed to a certain
dose of X-irradiation (about 5 krad for tobacco protoplasts),
plated in solidified medium at optimal density and over-
layered with sparsely plated, nonirradiated protoplasts. The
later protoplasts are then capable of division even when plated
at a very low density (e.g. 100 cells·ml^{-1}). Further tests
(Aviv and Galun, unpublished) indicated that X-irradiated
tobacco protoplasts had almost normal rates of precursor incorporation into protein and RNA. Moreover, even thymidine
incorporation into the DNase hydrolyzable fraction was only
slightly reduced in these protoplasts during the first few
days of culture. We thus concluded that although X-irradiation
impaired the ability of protoplasts to produce cell colonies
they obviously did not undergo total destruction. We
speculated that the replication ability of organelles in
X-irradiated protoplasts was not abolished.

We consequently based our first "donor-recipient" fusion
experiments (Zelcer *et al.* 1978) on the above consideration.
Our rationale was that after fusion of untreated *N.sylvestris*
("recipient") protoplasts with X-irradiated ("donor") protoplasts (having "cytoplasmic" controlled traits which differ
from those of *N. sylvestris*) all somatic hybrids (cybrids)
should have *N. sylvestris* nuclei and at least some of these
should have cytoplasmic traits of the "donor"

The "donor" protoplasts were obtained from the CMS tobacco
line (L-92) which was reported to have originated from the

cross *Nicotiana suaveolens* ♀ x *N. tabacum* ♂ and subsequently
repeatedly back-crossed to *N. tabacum* as pollinator.   The
morphology of L-92 is identical to *N. tabacum* (cv Xanthi)
but it is CMS (having stigmoid-petaloid anthers and no viable
pollen).   Due to uniparental transmission we assumed that the
chloroplasts and the mitochondria of L-92 are of *N. suaveolens*
(latter we could not confirm whether *N.suaveolens* or another
wild *Nicotiana* species was actually the original maternal
progenitor of L-92, thus in all subsequent references we preferred to term the organelles of L-92 as <u>L-92</u>, rather than as
sua).   The first fusion experiment can thus be written as:

$$syl/tbc/fert + tbc/L\text{-}92/ster \nearrow$$

meaning: *N.sylvestris* nucleus / *N. tabacum* chloroplasts/ male-
fertile plus *N. tabacum* nuclei/L-92 chloroplasts/male sterile
(the "lightning arrow" denotes X-irradiation).

It was anticipated that the fusion product would have a
selective advantage because the X-irradiated donors should not
produce colonies and the recipient (*N. sylvestris*) protoplasts
usually do not divide in our standard, mannitol containing,
medium.   The results of this experiment have been reported in
detail (Zelcer *et al.*, 1978; Galun and Aviv, 1979; Aviv and
Galun, 1979; Aviv *et al.*, 1980) and are described here only
briefly.   Most of the regenerated plants (20 plants from 5
different calli), termed Type A, had *N. sylvestris* morphology
but were male sterile, like the donor plants.   Moreover by
three tests - tentoxin sensitivity, electrofocussing of the LS
RUBPCase polypeptides and chl DNA restriction patterns - the
chloroplasts of Type A plants were shown to be identical to
those of L-92.   The uniformity of Type A plants in respect to
chloroplast composition and CMS was confirmed by out-crossing
to *N. sylvestris* and *N. tabacum*, as well as by androgenesis.
Type A plants can thus be regarded as *N. sylvestris* plants in
which the chloroplasts and the male fertility agents were ex-
changed with those of L-92.   For brevity Type A plants are
denoted   *syl/L-92/ster*.

We then planned another experiment in which Type A protoplasts served as recipients and *N. tabacum* cv Xanthi protoplasts were the donors: *syl/L-92/ster + tbc/tbc/fert* $\nearrow$. This
experiment was intended to furnish answers to the following
questions:
a.   Can the fertility be restored?
b.   Is the restored fertility correlated with *L-92* chloroplasts?

An affirmative answer to the first question would indicate
the general applicability of the "donor-recipient" system to
transfer undirectionally a cytoplasmic controlled trait.   A
negative answer to the second question would mean that male-
sterility control resides outside the chloroplasts (possibly
in the mitochondrial genome).

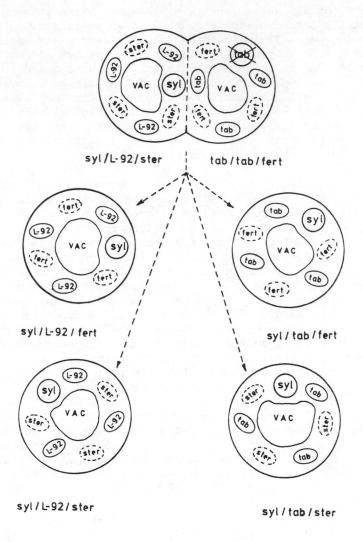

Fig. 1. Schematic presentation of the fusion experiment:
Type-A *(syl/L-92/ster)*+ *N. tabacum (tbc/tbc/fert)*.Four groups
of cybrid plants all having *N. sylvestris* nuclei but with
different chloroplast compositions and male fertility/sterility
were obtained. Circles-nuclei,solid-line elipsoids-chloroplasts;
broken line elipsoids - male fertility/sterility controlling
elements (probably mitochondria).

It would also hint at an independent assortment, after fusion, between chloroplasts and the male sterility/fertility controlling agents (i.e. probably the mitochondria). Actually such answers were obtained (Aviv and Galun, 1980). The results of one such fusion experiment (Fig. 1) can be summarized in the following list of cybrids:

$$7 \text{ plants from 3 calli:} \quad syl/tbc/ster$$
$$19 \text{ plants from 8 calli:} \quad syl/tbc/fert$$
$$8 \text{ plants from 4 calli:} \quad syl/L\text{-}92/fert$$

In some cases, plants with different chloroplast compositions were obtained from the same callus (e.g. *syl/tbc/fert* and *syl/L-92/fert*). Moreover a *syl/L-92/ster* plant (not reported in Aviv and Galun, 1980) was regenerated from the same callus which produced also *syl/tbc/ster* plants, indicating that sorting-out was still in process at the callus level.

As mentioned above, recent studies by Maliga and collaborators (e.g. Menczel *et al.*1981b; Sidorov *et al.*, 1981) augmented and confirmed the applicability of the "donor-recipient" fusion method to unidirectionally transfer organelles.

ASSORTMENT OF PLASTIDS AND MITOCHONDRIA FOLLOWING CYTOPLASMIC HYBRIDIZATION

The sorting-out of chloroplasts in hybrid and cybrid plants resulting from protoplast fusion was amply exemplified above. On the other hand there are also documented cases of presistent variegation, involving mutated chloroplasts, which are transmitted by sexual propagation. This later phenomenon is best explained by a "stable" heteroplastomic state which is carried beyond the egg cell. Can such a persistent heteroplastomic condition be achieved also by protoplast fusion? An affirmative answer would enable another very interesting question to be posed: Is there chl DNA recombination in angiosperms and if recombination does occur is it facilitated by prolonged coexistence of two different chloroplast genomes in the same cell?

Recent data from our laboratory (Dvora Aviv and Robert Fluhr, unpublished) show that certain protoplast fusions, of the donor-recipient type, result in two types of cybrids: homoplastomic and heteroplastomic plants. In one such fusion experiment *N. sylvestris* protoplasts (recipient) were fused with X-irradiated *N. tabacum* SR-1 protoplasts (donor). The two parental types are sensitive and resistant, respectively, to streptomycin (see Medgyesy *et al.*, 1980). Thirty eight cybrids (all with *N. sylvestris* nuclei) were regenerated. Of these 8 plants were homoplastomic and streptomycin resistant and their progeny did not segregate;

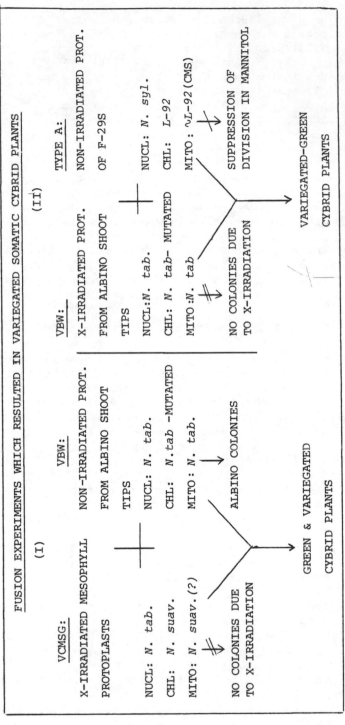

Fig. 2. – Schemes of two fusion experiments which resulted in variegated cybrid plants. VCMSG – a CMS tobacco line with *N. suaveolens* cytoplasm; VBW – vegatively propagated albino sector of a variegated tobacco line having mutated chloroplasts. (Unpublished data of D. Aviv and R. Fluhr.)

16 plants were homoplastomic and streptomycin sensitive; and
14 plants were heteroplastomic and their progeny segregated
to resistant and sensitive plants.

Other fusion tests, involving recipient or donor protoplasts with defective chloroplasts (i.e. plastome controlled
albinism) also resulted in cybrids with delayed sorting out of
chloroplasts. Fig. 2 presents two examples where the manifestation of this delay was persistent variegation. We may
thus conclude that chloroplast sorting out is common but not
exclusive.

We shall finally look at mitochondrial assortment and its
relation to chloroplast assortment. The information on this
subject is presently based mainly on the results of the fusion
of Type-A protoplasts with X-irradiated *N. tabacum* protoplasts
(*syl/L-92/ster + tbc/tbc/fert*) which was mentioned above in
respect to the lack of correlation between male sterility/
fertility and the chloroplast genome. The plant material
secured by this experiment led us to ask three questions:
(a) Is there sorting-out of mitochondria in cybrids resulting
from protoplast fusion? (b) Is there a correlation between
mitochondrial genomes and male sterility/ fertility? (c) Is
there an independent assortment of chloroplasts and mitochondria in cybrids resulting from parental types which differ
in respect to both organelles (Arzee-Gonen, 1980; Galun *et al*.
1981)? Restriction patterns, after fragmention with two endonucleases (Xho I and Sal I) served to characterize the mtDNA.
We first found that the two parental types had clearly
different mtDNAs. Restriction patterns following Sal I fragmentation showed that the parents shared only 13 of ∿ 30 fragments. Similar results were obtained after Xho 1 fragmentation.
Sorting out was apparently completed in all the cybrids; when
the mtDNA was analysed in the immediate cybrid plants or in
the maternal progenies of these cybrids the same patterns were
observed. Moreover, even cell suspension cultures of cybrids
maintained for prolonged periods showed the mtDNA restriction
patterns of the cybrids from which they were derived. There
were actually indications that sorting-out of mitochondria was
completed before chloroplast sorting-out. This possibility
emerges from the observation (Galun and Aviv, 1981) that from
some calli, resulting from this fusion, cybrids with *L-92*
chloroplasts as well as cybrids with *tbc* chloroplasts were
regenerated, while no such segregation from individual calli
was recorded in respect to mitochondria. Mitochondrial DNA
of cybrids rarely had restriction patterns identical to either
of their fusion parents (in conformation with Belliard *et al*.,
1979; Nagy *et al*., 1981) but the patterns were obviously
similar to one or the other of the parents. Furthermore, the

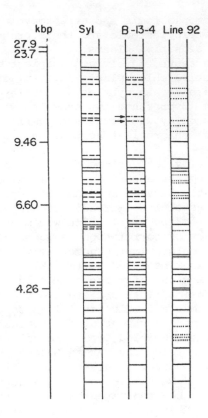

Fig. 3. mtDNA restriction pattern of the cybrid B-13-4(*syl/ tbc/fert*) compared to its two reference types: *N.sylvestris* (*syl/tbc/fert*) and *L-92 (tbc/L-92/ster)*. Xho 1 fragmentation; kbp-apparent kilobase pair length, horizontal solid lines- fragments common to both reference types; broken lines- fragments unique to *N. sylvestris*, dotted lines-fragments unique to L-92; arrows mark new fragments, not found in the reference plants (unpublished data of P. Arzee-Gonen).

mt DNA restriction patterns of all the male sterile cybrids were identical or similar to the respective pattern of the sterile (recipient) parent while all the male fertile cybrid's mt DNA restriction patterns were similar to those of the fertile (donor) parent (Fig. 3). Hence an obvious (but not proven as causal) correlation between male sterility/fertility and mitochondrial composition was established. Finally, we found all the four possible mitochondrial - chloroplast combinations among the cybrids:

$$syl/L-92/L-92^* \ (ster)$$
$$syl/L-92/tbc^* \ (fert)$$
$$syl/tbc/L-92^* \ (ster)$$
$$syl/tbc/tbc^* \quad (fert)$$

(The asterisks denote either similarity or identity with the respective mtDNA).

We, therefore, concluded that there was an independent assortment of chloroplasts and mitochondria in the cybrids resulting from this fusion experiment.

PERSPECTIVES

Many of the perspectives of cytoplasmic exchange by proto-plast fusion were hinted above. The main goals which may be achieved by this hybridization will be mentioned now only in brief. Several problems of basic plant cell genetics can now be approached. One such problem is the question of organelle sorting out - is it random or are there control mechanisms which facilitate or retard quick organelle assortment? The question of mtDNA and chl DNA recombination is now ready to be experimentally approached, though additional genetic markers, especially in mitochondria, are desirable. There are still many open questions regarding the reestablishment of division and differentiation capabilities of cells. Fusion between cells which have lost these abilities and cells which have retained them may furnish interesting answers. Finally, the role of organelles in reestablishing differentiation and/ or cell division may be revealed.

The possibility to unidirectionally transfer mitochondria or chloroplasts from donor to recipient plants has great promise in plant breeding. There are several known mito-chondria genome-controlled traits (e.g. male sterility, resistance to phytotoxins) which have breeding relevance. Likewise, several chloroplast genome characters are important breeding traits (e.g. pesticide resistances). The main obstacle in utilizing cytoplasmic hybridization in plant breeding is the small number of crop plants in which proto-plast culture and plant regeneration is presently possible.

ACKNOWLEDGMENTS

I am grateful to my collaborators Pazia Arzee-Gonen, Dvora Aviv, Marvin Edelman and Robert Fluhr, for permission to include their unpublished work in this article.

REFERENCES

Aviv, D. and Galun, E. (1980). *Theor. Appl. Genet*. 58,121.
Aviv, D., Fluhr, R., Edelman, M. and Galun, E. (1980). *Theor. Appl. Genet*. 56, 145.
Arzee-Gonen, P. (1980). Isolation and characterization of mitochondrial DNA in *Nicotiana* and the utilization of mtDNA restriction patterns as cytoplasmic markers in *Nicotiana* somatic hybrids. M.Sc. thesis, Submitted to the Feinberg Graduate School of the Weizmann Institute of Science, Rehovot.
Belliard, G., Vedel, F.,and Pelletier,G.(1979). *Nature* 281,401.
Belliard, G., Pelletier, G., Vedel, F.,and Quetier,F.(1978). *Molec. Gen. Genet*. 165, 231.
Chen, K., Wildman, S.G.,and Smith,H.H.(1977). *Proc. Natl. Acad. Sci. USA*. 74, 5109.
Fluhr,R.,and Edelman,M.(1981).*Mol. Gen. Genet*. 181, 484.
Galun, E. and Aviv, D. (1978). *IAPTC Newsletter* 25, 2.
Galun, E. and Aviv, D. (1979). *Monografia Genetica Agraria* 4, 153.
Galun, E. In "Methods in Chloroplast Molecular Biology" (M. Edelman, R.G. Hallick, N.-H. Chua eds) Elsevier/North Holland Biochemical Press, Amsterdam, (1982 in press).
Galun, E. and Raveh, D. (1975). *Radiation Bot*. 15, 79.
Gleba, Y.Y. In "Plant Cell and Tissue Culture - Principles and Applications" (R. Sharp, P.O. Larson, E.F. Paddock and V. Raghavan, eds) p. 774. Ohio State Univ. Press, Columbus, (1979).
Iwai, S., Nagao, T., Nakata, K., Kawashima, N. and Matsuyama,S. (1980). *Planta* 147, 414.
Izhar, S. and Power, J.B. (1979). *Plant Sci. Lett*. 14, 49.
Izhar, S. and Tabib, Y. (1980). *Theor. Appl. Genet*. 57, 241.
Kung, S.D., Gray, J.C., Wildman, S.G. and Carlson, P.S. (1975). *Science* 187, 353.
Medgyesy, P., Menczel, L. and Maliga, P., (1980). *Molec. Gen. Genet*. 1979, 693.
Melchers, G. In "Perspectives in Plant Cell and Tissue Culture" (I.K. Vasil ed), Inter. Rev. Cytol. Supp. 11B p. 241. Academic Press, N.Y., London, Toronto, Sydney, San Francisco, (1980).
Melchers, G., Sacristan, M.D. and Holder, A.A. (1978).*Carlsberg Res. Commun*. 43, 203.
Menczel, L., Galiba, G., Nagy, F. and Maliga, P. (1981a) *Genetics* (in press).
Menczel, L., Lazar, G. and Maliga, P. (1978). *Planta* 143,29.
Menczel,L., Nagy, F., Kiss, Zs.R. and Maliga, P. (1981b). *Theor. Gen. Genet*. 59, 191.

Nagy, G., Torok, I. and Maliga, P. (1981). Extensive rearrangements in the mitochondrial DNA in somatic hybrids of *Nicotiana tabacum* and *Nicotiana knightiana*. *Molec. Gen. Genet.* (in press)

Raveh, D., E. Huberman and E. Galun. (1973). *In Vitro*.9, 216.

Schieder, O. and Vasil, I.K. In "Perspectives in Plant Cell and Tissue Culture".(I.K. Vasil, ed) Intern. Rev. Cytol. Supp. 11B p. 21. Academic Press, N.Y., London, Toronto, Sydney, San Francisco (1980).

Scowcroft, W.R. and Larkin, P.G. (1981). *Theor. Appl. Genet.* 60, 179.

Sidorov, V.A., Menczel, L., Nagy, F. and Maliga, P. (1981). *Planta* 152, 341.

Wettstein, D. von Poulsen, C. and Holder, A.A. (1978). *Theor. Appl. Genet.* 53, 193.

Zelcer, A., Aviv, D. and Galun, E. (1978). *Z. Pflanzenphysiol.* 90, 397.

CELL CULTURE MUTANTS AND
THEIR USES

Pål Maliga, Låszló Menczęl, Vladimir
Sidorov[1], Låszló Mårton, Ágnes Cséplĕ,
Pěter Medgyesy, Trinh Manh Dung[2],
Gabriella Låzår, Ferenc Nagy

Institute of Plant Physiology
Biological Research Center
Hungarian Academy of Sciences
Szeged, Hungary

I. INTRODUCTION

The use of plant tissue cultures in studies on
mutation and selection is attractive because one
can manipulate in a confined place a large population of haploid or diploid cells which can be regenerated into whole plants. The potentials of the
plant system were described by Melchers and Bergman
as early as 1958. Lack of routine methods of cell
culture, and plant regeneration delayed realization
of such projects until the early 1970's. Research
since then has been stimulated by the need for
marker mutations for cellular genetic manipulations,
and by the potential for use of selected lines in
physiological studies and breeding. Progress in the
selection and characterization of cell lines in
plant tissue culture has been reviewed recently by
Widholm (1977a) and Maliga (1980). Separate reviews
on amino acid analog-resistant lines (Widholm 1977b),
metabolic and chlorophyll-deficient mutants

[1]Present address: Institute of Botany, Kiev, U.S.S.R.
[2]Present address: Institute of Experimental Botany,
            Ho Chi Minh City, Vietnam

221

(Schieder, 1978), resistance mutants (Maliga, 1978) and on aspects relevant to application in breeding (Maliga, 1981) have also been published. In this paper therefore only important recent findings will be mentioned which were not covered in the cited reviews (Section II). In Section III experiments will be described from this laboratory to demonstrate the suitability of *Nicotiana plumbaginifolia* as a model species for cellular genetic studies. At the end of the paper present trends will be analysed, and the problem of model species discussed (Section IV).

II. RECENT PROGRESS IN MUTANT SELECTION AND CHARACTERIZATION

A. *Auxotrophic Cell Lines*

Application of the methods of molecular biology to flowering plants, e.g. transformation by isolated DNA, would be greatly facilitated by the availability of characterized auxotrophic mutants. Reports on such mutants are rare (reviews in Introduction). There has recently however been fast progress since selection of auxotrophic cell lines is carried out in cultures of true haploid cells, and not in dihaploid cells like those of *Nicotiana tabacum,* or in cultures which were initially haploid but have become polyploid during extended cultivation.

1. *Nitrate Reductase-Deficiency.* Selection for resistance to chlorate, an analog of nitrate, resulted in the isolation of partially or fully nitrate reductase-deficient (NR$^-$) lines in *Nicotiana tabacum* (Müller and Grafe, 1978). The NR$^-$ cells are unable to utilize nitrate, and growth is dependent on the availability of reduced nitrogen. The efficiency of selection may explain that fully deficient cells could be obtained in dihaploid cells since in that case theoretically two independent mutations are required to obtain non-leaky auxotrophs. By crossing resistant regenerates it could be shown that this indeed was the case (A.J.Müller, pers. comm.). The *N.tabacum* lines have been characterized biochemically (Mendel and Müller, 1979; 1980; Mendel et al., 1981). Similar cell lines have been

isolated and characterized in *Datura innoxia* (King and Khanna, 1980), *Rosa damascena* (Murphy and Imbrie, 1981), *Hyoscyamus muticus* (Strauss et al., 1981) and *Nicotiana plumbaginifolia* (Section III, A.*1*).

By screening seedlings mutants in nitrate reductase were obtained in *Arabidopsis thaliana* (Oostindier-Braaksma and Feenstra, 1973), *Hordeum vulgare* (Kleinhofs et al., 1980; Kuo et al., 1981) and in *Pisum sativum* (Feenstra and Jakobsen, 1980). The mutant plants contained at least a low level of nitrate reducing activity.

2. *Other Types of Auxotrophic Lines*.   By individual testing of calli grown from mutagenized haploid cells lines requiring amino acids, nucleic acid bases and vitamins have been recovered at a frequency of 10^{-3}-10^{-4}. In *Datura innoxia* a pantothenate and an adenine auxotrophic line (Savage et al., 1979; King et al., 1980), and in soybean a line requiring asparagine or glutamine (K.G.Lark, pers.comm.) were obtained. In *Hyoscyamus muticus* histidine (2 lines) tryptophan, and nicotinic acid (3 lines) auxotrophs (Gebhardt et al., 1981), and in *Nicotiana plumbaginifolia* isoleucine, uracil and leucine requiring lines (Sidorov et al., 1981a; Section III, A,*1*) were isolated using the same method.

After arsenate treatment, which selectively kills dividing plant cells (Polucco, 1979), a line with a requirement for cysteine+methionine+lysine+threonine was obtained in *Datura* (J.King, pers. comm.).

B. *Amino Acid and Amino Acid Analog Resistance*

Stimulated by the hope of producing plants with increased amounts of certain nutritionally important amino acids, efforts have been made to select for resistance to amino acids, and their analogs (reviewed by Widholm, 1977a,b).

Recently, selection of a lysine+threonine (LT) resistant line was described from tissue culture of maize (Hibberd et al., 1980). Analysis of free amino acids showed substantial increase in methionine and threonine. Plants have been regenerated from another LT resistant culture *LT19* (Hibberd

and Green, 1981). Resistance was shown to be inherited as a single dominant nuclear character.
Analysis of free amino acids in kernels homozygous
for $Ltr^{x}-19$ indicated that threonine was increased
75-100 times. Free threonine overproduction increased the total threonine by 33-59%. A similar
LT resistant line has been obtained in barley by
Bright et al. (1981) by screening M2 embryos. Resistance in the $R2501$ line was conferred by a dominant gene $Lt1$. No homozygous $Lt1/Lt1$ fertile
plants have been recovered. Threonine in the
soluble fraction of mature seeds was increased
from 0.8 to 9.6% of the total threonine content.

   Lysine overproducing (10-15 x) plants were
obtained from a *Nicotiana sylvestris* cell line
which was selected for resistance to aminoethylcysteine (AEC), an analog of lysine. AEC resistance was seed-transmissable (I.Negrutiu and M.
Jacobs, pers.comm.). In rice regenerates, after
selection for resistance to AEC, no specific increase was found in lysine content (Schaeffer and
Sharpe, 1981; Schaeffer, 1981). The role of AEC
selection in the increase in the amount of seed
protein in the same plants needs to be clarified.

   In leaves of barley plants selected for resistance to 4-hydroxy-L-proline the content of
soluble proline was higher by a factor of up to
six-fold. Resistance was due to a partially dominant nuclear gene (Kueh and Bright, 1981).

C. *Temperature-Sensitive Variants Impaired in
   Somatic Embryogenesis*

   In carrot, plant regeneration takes place via
embryogenesis. Large number of somatic embryos produced allowed the recovery of temperature-sensitive
variants impaired in early embryogenesis (A.M.
Breton and Z.R.Sung, pers.comm.). The first class
is blocked before any growth in embryogenic medium
can take place; the second class can grow but does
not from any embryos, and the third is blocked in
the first stage of embryo development, i.e. the
globular embryo. In a different laboratory similar
carrot lines were obtained (M.Terzi, G.Guillano,
F.Lo Schiavo, pers.comm.).

D. *Salt Tolerance*

Tolerance to salt toxicity is a problem with increasing practical importance since irrigation leads to salt accumulation. It is of interest therefore that seed progeny of plants regenerated from NaCl resistant *Nicotiana tabacum* cell lines was reportedly more tolerant to irrigation by water containing NaCl (Nabors et al., 1980). In a *Nicotiana sylvestris* NaCl-resistant line proline accumulation was excluded as the mechanism of tolerance (Dix and Pearce, 1981). Mercuric chloride resistance described in *Petunia* (Colijn et al., 1979) was retained after regeneration into plants (Kool and Colijn, 1981).

E. *Characterization of the Selected Lines*

In cases in which plant regeneration is not possible, or the plants are not fertile, somatic hybridization has been employed to characterize the selected lines.

In *Nicotiana sylvestris* somatic hybrids aminoethylcysteine and 5-methyltryptophan resistances were shown to be dominant and semidominant traits, respectively (White and Vasil, 1979). The same two resistances were found to be dominant in carrot somatic hybrids (Harms et al., 1981). Cycloheximide resistance in carrot is a function expressed in differentiated plant tissues (Sung et al., 1981). A variant, WCH105, was isolated which expresses the resistance in the callus as well as in regenerated plantlets. The green WCH105 protoplasts were fused with albino carrot. Resistance was recessive in the somatic hybrids. The combined recessive and resistant phenotype of this trait allowed the recovery of resistant segregants from the sensitive hybrids at a frequency of 10^{-4}, 1000 times higher than the spontaneous frequency of resistance (Lázár et al., 1981). Griseofulvin-induced chromosome segregation may be a useful tool in future studies of similar somatic hybrids (Lo Schiavo et al., 1980).

Localization of a cytoplasmic streptomycin resistance factor in *Nicotiana* was not possible in sexual crosses since both potential carriers of the genetic information, chloroplasts and mitochondria, are inherited maternally. Study of plastid segregation in somatic hybrids of *Nicotiana tabacum*

(streptomycin resistant) and *Nicotiana knightiana*
(sensitive) has led to the identification of chloroplasts as carriers of the resistance factor
(Menczel et al., 1981).

A different approach to characterize mutations,
study of a second site revertant, has been employed
in case of a temperature-sensitive mutant, <u>ts</u> 4, of
Nicotiana tabacum. The analysis of the two inter-
acting mutations and their effects on the same
processes allowed deductions to be made about the
normal regulatory interactions of the effected gene
products (Malmberg, 1980). Unselected traits, tricotyly of seedlings and hydroxyurea resistance,
were detected in picloram resistant *Nicotiana tabacum* mutants (Chaleff and Keil, 1981). Tricotyly
was thought to be a physiological response to
passage through cell culture. Hydroxyurea resistance was inherited as a single dominant nuclear
mutation.

III. *NICOTIANA PLUMBAGINIFOLIA* – A PROSPECTIVE MODEL SPECIES

Widespread use of one model species for mutation induction in cultured cells would greatly
stimulate genetic research with plant cells and
the exchange of information between the fields of
plant breeding, genetics and molecular biology.
Desirable properties of a model plant species are:
fast growth in culture, maintained plant regeneration ability, easy protoplast culture, low chromosome number, true haploid genome, availability of
haploid plants and of a genetic map. *Nicotiana
plumbaginifolia* (2n=2X=20), chosen on the basis of
the survey of *Nicotiana* species by Bourgin et al.
(1979), is widely used in our laboratory. This plant
satisfies all criteria of a model species except
for the existence of a genetic map. Suitability of
the species will be shown for mutant selection
by the isolation of auxotrophic, antibiotic resistance and pigment mutations (Section III.A) and
for fusion by experiments on chloroplast transfer
(Section III.B). Alternative model species will be
mentioned in Section IV.

A. *Mutant Selection in* Nicotiana plumbaginifolia

   1. Auxotrophic Cell Lines. Haploid protoplasts
were prepared from leaves of *N. plumbaginifolia*
plants, and mutagenized by irradiation. The efficiency of mutagenic treatment was indicated by the
increased frequency of pigment mutants in the same
experiment.(Table 1). Protoplasts were grown to
colonies in complete medium, then individually
tested for nutritional requirement in minimal medium (for details see footnote of Table 1). Three
out of 14,229 colonies, tested after mutagenesis
were unable to grow on minimal medium (Table 1).
These three calli were later identified as a uracil,
an isoleucine, and a leucine requiring line by
growing them on media supplemented with various
amino acids, vitamins and nucleic acid bases

*Table 1. The frequency of auxotrophic and pigment
        deficient clones[a]*

Dose[b] $(J\ kg^{-1})$	Survival[c] (%)	Number (frequency) of		
		Clones tested	Auxotro- phics	Pigment deficients
13	57	3727	0	3 (8×10^{-4})
16	23	3705	0	8 (2×10^{-3})
19	15	2377	0	7 (3×10^{-3})
23	11	4420	3	36 (8×10^{-3})

[a]*Protoplasts were cultured in K3 medium (Nagy and
Maliga, 1976) supplemented with 0.1% casein hydrolysate and 0.05% yeast extract (complete medium).
The colonies were then individually transferred onto RM medium (Linsmaier and Skoog, 1965) containing 0.1 mg l^{-1} 1-naphtaleneacetic acid and 1.0 mg
l^{-1} benzyladenine (RMOP medium) and the same supplements as before. This medium induces greening, so
pigment deficient clones could be identified at
this step. Green calli were then subcultured onto
minimal (casein hydrolysate and yeast extract o-
mitted) RMOP medium to identify auxotrophics.*
[b]*Dose rate of the ^{60}Co radiation was 0.042 J kg^{-1}
sec^{-1}.*

[c]*Percent of non-irradiated control. Plating efficiency in cultures of non-irradiated protoplasts
was around 70%.*

following the test system of Holliday (1956).
Diploid plants have been regenerated from the isoleucine requiring line, ILE401. The plants, and the
calli initiated from them, have an absolute requirement for isoleucine. The leucine-requiring
calli have also formed shoots on medium supplemented with leucine so we hope to obtain flowers
and make crosses with this line. The uracil auxotrophic line, URA401, did not regenerate shoots
under any conditions tried. URA401 protoplasts were
fused therefore with haploid protoplasts of a recessive nuclear pigment mutant A28 (see III.A.2),
and intraspecific hybrids were selected as green,
prototrophic calli. Morphogenic ability in the somatic hybrids was expressed, as described in a previous case after fusing non-morphogenic cell types
(Maliga et al., 1977). Plants have been transferred
to the greenhouse in order to test sexual transmission of the auxotrophic trait.

     In the ILE401 line auxotrophy was shown to be
due to deficiency in threonine deaminase. The other
two lines have not been characterized biochemically.
Some of these data have been published (Maliga et
al., 1981; Sidorov et al., 1981b).

     2. Pigment-Deficient Lines.   Increase in the
frequency of pigment-deficient colonies was used to
test the efficiency of mutagenic treatment in the
auxotrophic isolation experiment (Table 1). Since
regenerates with haploid (X=10) chromosome numbers
were obtained, these plants were further characterized in order to use them for rescueing auxotrophic mutations as described in Section III.A.1.
The pigment-deficient plants cannot be grown in the
greenhouse, and many of them failed to survive at
high light intensities even if grafted onto a proper rootstock. Protoplasts prepared from the leaves
of five of these plants therefore were fused in
pairvise combinations, and the mutations were shown
by the formation of green colonies to be non-allelic
(Table 2). Green regenerates have been transferred
to the greenhouse for testing segregation after
self pollination.
     An albino line, A25, was obtained after ^{60}Co
irradiation in a diploid protoplast culture of
N.plumbaginifolia. Fusion of A25 protoplasts with
N.tabacum SR1-A15 protoplasts resulted in the
formation of green somatic hybrids. Pigment deficiency in the SR1-A15 mutant is maternally inherited

*Table 2. Fusion-Complementation of Pigment
          Deficiency*

Ploidy	Fusion combinations	Comple-mentation	Plant
	A28 + A27	+	+
	A28 + X32	+	+
n	X32 + A27	+	+
	X32 + A25	+	+
	A28 + A25	+	+
2n	A25 + SR1-A15	+	+

(unpublished). Complementation therefore may also
be explained by a recessive nuclear pigment mutation in the A25 line.

   3. *Nitrate Reductase-Deficient lines.* Nitrate
reductase (NR)-deficient cell lines were isolated
in cultures of mutagenized haploid protoplasts by
their resistance to 40 mM NaClO3, essentially as
described by Müller and Grafe (1978). Properties of
the NR⁻ deficient lines, as compared to wild-type
N.plumbaginifolia (NP) cultures, are summarized in
Table 3.
   Most of the lines studied, except two, contained no detectable NR activity. Plant regeneration
was possible only from those two, NA8 and NA14,
which had some residual NR activity. Formation of
functional plants therefore seems to be dependent
on the maintenance of some residual nitrate reducing
capacity. Xanthine dehydrogenase (XDH) activity was
detected in some lines (NA series) but not in
others (NX series). The lines with the XDH activity
are likely to be impaired in the NR apoprotein
whereas those with no XDH activity are probably defective in the molybdenum-cofactor biosynthesis
(Mendel and Müller, 1978). NR activity in two of
the lines, NX1 and NX9 could be partially restored
by increasing the molybdenum concentration in the
culture medium. These lines therefore seem to be
similar to *cnx* lines in *N.tabacum* (Mendel et al.,
1981). Lines NX21 and NX24 do not respond to molybdenum, and are new types which have not been described in flowering plants.

Table 3. Properties of Nitrate Reductase-Deficient
Lines

Line	NR activity	XDH activity	No response	Plant
Np	100%	+	−	+
NA8	0-7%[a]	+	−	+
NA14	1-6%[a]	+	−	+
NA1	−	+	−	−
NA9	−	+	−	−
NA2	−	+	−	−
NA14	−	+	−	−
NA36	−	+	−	−
NX21	−	−	−	−
NX24	−	−	−	−
NX1	−	−	+	−
NX9	−	−	+	−

[a]*Activity depends on culture age*

By *in vitro* enzyme complementation of NA and
NX lines NR activity could be restored (unpublished
results of L.Márton and R.Mendel). There is no enzyme complementation between the two types of NX
lines. Preliminary data on somatic hybrids indicate
that the two types of NX lines are non-allelic.

  4. Streptomycin and Lincomycin Resistant Lines.
Streptomycin resistant lines were selected in
diploid leaf protoplast cultures by their ability
to green in K3 medium (with growth regulators as in
RMOP, given in Table 1) containing 1 mg ml^{-1} streptomycin sulfate. Spontaneous and induced frequency
of streptomycin resistance was 2.7×10^{-6} (2 resis-
tants among 7.3×10^5 calli) and 87×10^{-6} (10 resis-
tants among 1.2×10^5 calli after treatment with
0.1 mM N-ethyl-N-nitrosourea), respectively. Resistance was transmitted maternally to F1 in two
clones.
  Lincomycin resistant lines were selected the
same way on a medium containing 1 mg ml^{-1} lincomycin hydrochloride. Maternal inheritance of the resistance was shown in one line.

B. *Transfer of Mutant Plastids into* Nicotiana plumbaginifolia

Experiments aimed at transferring chloroplasts from *N.tabacum* into *N.plumbaginifolia*, without the transfer of nuclear genetic material, will be described in this Section. As the method of transfer protoplast fusion has been employed. Transfer by protoplast fusion can be obtained only in cases in which the donor nucleus is eliminated in the primary interspecific fusion product. Organelles also segregate in the progeny of such heterokaryons which may result in the formation of new nucleus-organelle combinations, that is organelle transfer.

As plastid donor two *N.tabacum* mutants were used: SR1, which carries a plastome-coded streptomycin resistance factor (Menczel et al., 1981), and a light-sensitive plastome mutant (Wettstein and Erikson, 1965; Wildman et al., 1973).

1. Irradiation to Eliminate Donor Nuclei. Protoplast fusion is followed by the fusion of nuclei in the majority of heterokaryons. For the elimination of the donor nuclei in a high proportion of heterokaryons irradiation has been successfully applied (Zelcer et al., 1978; Aviv and Galun, 1980; Sidorov et al., 1981a). In these studies only one dose has been used. Experiments were carried out using different doses in order to establish the optimal conditions for chloroplast transfer. It has also been investigated whether the irradiated nucleus tends to be eliminated as a whole or a few chromosomes are normally retained from the irradiated cells. The protoplasts of the SR1 donor were irradiated before fusion with ^{60}Co γ-rays using doses higher that LD_{100}. Such high doses prevented colony formation from the resistant *N.tabacum* cells so heterokaryon derived clones containing SR1 plastids could be selected by their streptomycin resistance.

From the resistant clones (153 investigated) plants were regenerated and classified as *N.plumbaginifolia* or somatic hybrid. Using 0, 50, 120, 220 and 300 J kg^{-1} the percentage of clones giving *N. plumbaginifolia* regenerates was 1.4, 44, 57, 84 and 70, respectively. (In the absence of irradiation division of resistant parental protoplasts was prevented by iodoacetate treatment.) Irradiation of

the donor protoplasts therefore has led to efficient elimination of the donor nuclei. Chromosome
counts in 43 clones indicated that elimination has
been complete. Among the regenerates which were
classified as *N.plumbaginifolia* no aneuploids were
found to suggest retention of a few chromosomes
from the irradiated parent.

Resistance is inherited maternally (5 clones
tested). The presence of *N.tabacum* plastids in the
N.plumbaginifolia plants, that is chloroplast transfer, was confirmed by chloroplast DNA fragmentation
patterns after EcoRI digestion. Data mentioned in
this Section will be published in a paper by Menczel
et al. (1982).

2. *Cytoplast-Protoplast Fusion for Chloroplast
Transfer.* Cytoplasts are obtained by eliminating
the nuclei from the protoplasts by centrifugation
(Lörz et al., 1981). Cytoplasts, having no nucleus,
are not able to form colonies. They contain, however, chloroplasts (and mitochondria) and are an
ideal organelle donor if they also contain a selectable cytoplasmic trait. After protoplast-cyto-
plast fusion selection for a chloroplast trait
originally present in cytoplasts should lead to the
identification of clones with plastids from the
donor and the nucleus of the recipient.

Efficiency of transfer in this system depends
on the amount of contaminating donor nuclei (free
or in protoplasts) in the cytoplast preparation,
which may be 2-7% (Lörz et al., 1981). In order to
compare the cytoplast-protoplast system with other
methods of chloroplast transfer, cytoplasts were
prepared from callus protoplasts of the streptomycin
resistant tobacco mutant, SR1, and fused with protoplasts of sensitive *N.plumbaginifolia.*

Among 75 clones 16 (17%) were found in which
N.plumbaginifolia regenerates were obtained. In
one clone two different types *N.plumbaginifolia* and
somatic hybrid plants regenerated: for explanation
see Medgyesy et al., 1980. Resistant regenerates
were obtained in 8 clones and shown to contain SR1
plastids by the EcoRI restriction pattern of chloroplast DNA. Resistance in these clones was inherited maternally. Plants regenerated in 9 resistant clones turned out to be sensitive by the test
of leaf sections. This phenomenon is explained by
the maintenance of plastid heterogeneity in the
phenotypically resistant calli, and plastid segre-

gation on the non-selective plant regeneration
medium. Such segregants were obtained also in previous experiments (Menczel et al., 1982).

   The majority, 56 clones, regenerated somatic
(nuclear) hybrid plants. These hybrids originate
from fusion with contaminant SR1 protoplasts. Two
clones have *N.tabacum* nuclei and *N.plumbaginifolia*
cytoplasm. This should be the result of nuclear
segregation in heterokaryons (Medgyesy et al., 1980;
Menczel et al., 1981; Sidorov et al., 1981).

   The 10% (8 clones out of 75) overall transfer
efficiency of cytoplast-mediated chloroplast transfer is low as compared to other methods (see Section III.B.*1*). It is, however, sufficiently high
to be used in transfer experiments. The application
of this technique has the advantage that neither
the donor, nor the recipient, should be treated by
a toxic substance (iodoacetate), or subjected to a
treatment (irradiation) which is mutagenic (unpublished results of P.Maliga, H.Lörz, G.Lázár and
F.Nagy).

   3. Plastid Transfer Using No Selective Markers.
The applicability of the methods described in Section III.B.*1* and *2* requires the use of a selectable
plastome mutation. Using a scheme based on double-
inactivation the need for a selectable marker could
be eliminated (Sidorov et al., 1981a).

   Irradiation (60 J kg^{-1}) of the protoplasts of a
light-sensitive plastome mutant of *N.tabacum* (donor)
and iodoacetate treatment of *N.plumbaginifolia* (recipient) protoplasts prevented the division of the
parental cells. Metabolic complementation, however,
took place in the interspecific fusion products
which survived and formed calli. Irradiation in
this case also led to the elimination of the *N.ta-
bacum* nuclei in many fused protoplasts: The proportion of the clones which produced *N.plumbagini-
folia* regenerates was 38% (whereas it was 1.4% in
the absence of irradiation; Section III.B.*1*). Regenerates in most clones (in 31 out of 34) contained plastids from the irradiated parent. Irradiated plastids therefore seem to have an improved
competitive ability in this system.

IV. CONCLUDING REMARKS

Success in isolating auxotrophic cell lines,
and application of more sophisticated methods for
their characterization indicate that research in
this area is reaching a more mature phase. Genetic
and biochemical characterization is often carried
out both in cultured cells and after plant regeneration, fully exploiting the possibility to redif-
ferentiate plants from cultured cells. Production
of plants with improved amino acid composition, and
with tolerance or resistance to diseases, (for references see reviews in Introduction) confirm expectations concerning possible applications in
breeding.

Most success in the applied area has been obtained with corn and potato. Unfortunately, tissue
culture of these species, and of other crop plants
is not sufficiently refined for them to be used as
model species. *N. tabacum* which is still widely used
in basic studies needs to be replaced because of
its high chromosome number (2n=4X=48), and amphiploid origin. Possible candidates for replacement,
based on their proven suitability of mutation induction and somatic hybridization are *Datura innoxia* (Schieder, 1978; Schieder and Vasil, 1980;
Savage et al., 1979), *Hyoscyamus muticus* (Wernicke
et al., 1979; Gebhardt et al., 1981), *Petunia*
(Hanson and Kool, 1981) and *Daucus carota* (Sung and
Dudits, 1981). From among the diploid species in
the genus *Nicotiana* first *N. sylvestris* was considered. There has been success in somatic hybridization (e.g. White and Vasil, 1979) and mutant selection (e.g. Maliga, 1981). Initial reports on
protoplast culture of this species (Nagy and Maliga,
1976; Bourgin et al., 1976), however, were followed
by at least four other papers on the same subject
indicating difficulties in obtaining reproducibly
high frequencies of protoplast division. This
problem is not there with *N. plumbaginifolia* which
seems to be suitable for both mutation and fusion
studies (Section III).

It is hoped that the number of model species
currently in use for mutation induction studies
will be reduced. Establishment of a stock of characterized mutations in one (or a few) species
should be mutually beneficial for workers interested in genetic manipulation of cultured cells.

ACKNOWLEDGMENTS

We thank P.J.Dix, S.W.J.Bright, D.Dudits, C.E. Green, C.Gebhardt, M.Jacobs, A.J.Kool, J.King, K.G. Lark, H.Lörz, A.J.Müller, I.Negrutiu, G.W.Schaeffer and Z.R.Sung for making available their unpublished results for the review.

REFERENCES

Aviv,D., and Galun,E. (1980). *Theor.Appl.Genet.* *58*, 121.
Bourgin,J.P., Chupeau,Y., and Missioner,C. (1979). *Physiol.Plant.* *45*, 288.
Bright,S.W.J., Norbury,P.B., and Miflin,B.J. (1981). *Biochem.Genet.* (in the press)
Chaleff,R.S., and Keil,R.L. (1981). *Molec.Gen. Genet.* *181*, 254.
Colijn,C.M., Kool,A.J., and Nijkamp,H.J.J. (1979) *Theor.Appl.Genet.* *55*, 101.
Dix,P.J., and Pearce,R.S. (1981). *Z.Pflanzenphysiol.* *102*, 243.
Feenstra,W.J., and Jakobsen,E. (1980). *Theor.Appl. Genet.* *58*, 39.
Gebhardt,C., Schnebli,V., and King,P. (1981). *Planta,* *153*, 81.
Hanson,M.R., and Kool,A.J. (1981). *In* "Molecular Biology in the genus *Petunia*" (K.C.Sink, and J.R.Power, eds.) TAG monograph. Springer Verlag, Berlin (in the press)
Harms,C.T., Potrykus,I., and Widholm,J.M. (1981). *Z.Pflanzenphysiol.* *101*, 377.
Hibberd,K.A., and Green,C.E. (1981). *Proc.Natl. Acad.Sci.* USA (in the press)
Hibberd,K.A., Walter,T., Green,C.E., and Gegenbach,B.G. (1980). *Planta 148*, 183.
Holliday,R. (1956). *Nature 178*, 987.
King,J., and Khanna,V. (1980). *Plant Physiol. 66*, 632.
King,J., Horsch,R.B., and Savage,A.D. (1980). *Planta 149*, 480.
Kleinhofs,A., Kuo,T., and Warner,R.L. (1980). *Molec.Gen.Genet. 177*, 421.
Kool,A.J., and Colijn,C.M. (1981). *In* "Induced Mutations as a Tool for Crop Plant Improvement" (P.Howard-Kitto, ed.), p. 407. IAEA publ. No. 251, Vienna.

Kueh,J.S.H., and Bright,S.W.J. (1981). *Planta 153*, 166.

Kuo.T., Kleinhofs,A., Somers,D., and Warner,R.L. (1981). *Molec.Gen.Genet. 181*, 20.

Lázár,G.B., Dudits,D., and Sung,Z.R. (1981). *Genetics* (in the press)

Linsmaier,E.M., and Skoog,F. (1965). *Physiol. Plantarum 18*, 100.

Lörz,H., Paszkowski,J., Dierks-Ventling,C., and Potrykus,I. (1981). *Physiol.Plantarum 53*, 385.

Maliga,P. (1978). *In* "Frontiers of Plant Tissue Culture, 1978". (T.A.Thorpe, ed.), p. 381. Univ.of Calgary Press, Calgary, Alberta.

Maliga,P. (1980). *In* "Perspectives in Plant Cell and Tissue Culture" (I.K.Vasil, ed.), p. 225. Int.Rev.Cytol. suppl. 11A Academic Press, New York.

Maliga,P. (1981). *Theor.Appl.Genet. 60*, 1.

Maliga,P., Lázár,G., Joó,F., H.-Nagy,A., and Menczel,L. (1977). *Molec.Gen.Genet. 157*, 291.

Maliga,P., Sidorov,V., Cséplü,A., and Menczel,L. (1981). *In* "Induced Mutations as a Tool for Crop Plant Improvement" (P.Howard-Kitto, ed.), p. 339. IAEA publ. No. 251, Vienna.

Malmberg,R.L. (1980). *Cell 22*, 603.

Medgyesy,P., Menczel,L., and Maliga,P. (1980). *Molec.Gen.Genet. 179*, 693.

Melchers,G., and Bergmann,L. (1959). *Ber.Dtsch. Bot.Ges. 71*, 459.

Menczel,L., Nagy,F., Kiss,Zs., and Maliga,P. (1981). *Theor.Appl.Genet. 59*, 191.

Menczel,L., Galiba,G., Nagy,F., and Maliga,P. (1982). *Genetics* (in the press)

Mendel,R.R., and Müller,A.J. (1980). *Plant.Sci.Lett. 18*, 277.

Mendel,R.R., and Müller,A.J. (1978). *Molec.Gen. Genet. 161*, 77.

Mendel,R.R., and Müller,A.J. (1979). *Molec.Gen. Genet. 177*, 145.

Mendel,R.R., Alikulov,Z.A., Lvov,N.P., and Müller, A.J. (1981). *Molec.Gen.Genet. 181*, 395.

Murphy,T.M., and Imbrie,C.W. (1981). *Plant Physiol. 67*, 910.

Müller,A.J., and Grafe,R. (1978). *Molec.Gen.Genet. 161*, 67.

Nabors,M.W., Gibbs,S.E., Berstein,C., and Meis,M. (1980). *Z.Pflanzenphysiol. 97*, 13.

Nagy,J.I., and Maliga,P. (1976). *Z.Pflanzenphysiol.* *78*, 453.

Oostindier-Braksma,F.J., and Feenstra,W.J. (1973). *Mutation Res. 19*, 175.

Pollaco,J.C. (1979). *Planta 146*, 155.

Savage,A.D., King,J., and Gamborg,O.L. (1979). *Plant Sci.Lett. 16*, 376.

Schaeffer,G.W. (1981). *Environ.Exptl.Bot.* (in the press).

Schaeffer,G.W., and Sharpe,F.T. (1981). *In Vitro 17*, 345.

Schiavo,F., Nuti Ronchi,V., and Terzi,M. (1980). *Theor.Appl.Genet. 58*, 43.

Schieder,O. (1978). *In* "Frontiers of Plant Tissue Culture, 1978" (T.A.Thorpe, ed.), p. 393. Univ.of Calgary Press, Calgary, Alberta.

Schieder,O., and Vasil,I.K. (1980). *In* "Perspectives in Plant Cell and Tissue Culture" (I.K. Vasil, ed.), p. 21. Int.Rev.Cytol. suppl. 11B Academic Press, New York.

Sidorov,V.A., Menczel,L., Nagy,F., and Maliga,P. (1981a). *Planta 152*, 341.

Sidorov,V.A., Menczel,L., and Maliga,P. (1981b). *Nature 294*, 87.

Strauss,A., Bucher,F., and P.J.King (1981). *Planta 153*, 75.

Sung,Z.R., and Dudits,D. (1981). *In* "Genetic Engineering in the Plant Sciences" (Panopoulos, N.J., ed.), Praeger, New York (in the press).

Sung,Z.R., Lázár,G.B., and Dudits,D. (1981). *Plant Physiol. 68*, 261.

Wernicke,W., Lörz,H., and Thomas,H. (1979). *Plant Sci.Lett. 15*, 239.

Wettstein,D.V., and Erikson,G. (1965). *In* "Genetics Today" Proc.11th Int.Congr.Genet. (Geerts,S.J., ed.), Vol.3, p. 591. Pergamon Press, Oxford.

White,D.W.R., and Vasil,I.K. (1979). *Theor.Appl. Genet. 55*, 107.

Wildman,S.G., Lui-Liao,C., and Wong-Staal,F. (1973). *Planta 113*, 293.

Widholm,J.M. (1977a). *In* "Plant Tissue Culture and its Bio-technological Applications" (W.Barz, E.Reinhard, and M.J.Zenk, eds.), p. 112. Springer-Verlag, Berlin and New York.

Widholm,J.M. (1977b). *Crop.Sci. 17*, 597.

Zelcer,A., Aviv,D., and Galun,E. (1978). *Z.Pflanzenphysiol. 90*, 397.

CHAPTER 12

SOMATIC HYBRIDIZATION: A NEW METHOD FOR PLANT IMPROVEMENT

Otto Schieder

Max-Planck-Institut für Züchtungsforschung (Erwin-Baur-Institut)
5000 Köln 30, Federal Republic of Germany

I. INTRODUCTION

The development of viable embryos in most interspecific or intergeneric crosses is not achieved because of specific inhibition or prevention of key steps in pollination, pollen tube growth, fertilization, and embryo or endosperm development. A new tool for the genetic manipulation of plants become available with the development of methods for the enzymatic isolation of protoplasts and their subsequent regeneration to plants (Vasil and Vasil, 1980), and of methods which allow the fusion of protoplasts of genetically different lines or species (Keller and Melchers, 1973; Kao and Michayluk, 1974). Four major aspects of protoplast fusion should be considered: I. The production of fertile amphidiploid somatic hybrids of normally in compatible species. II. The production of heterozygous lines within one plant species which normally will be propagated only vegetatively, as for example in potato. III. The transfer of only a part of the nuclear genetic information from one species to another using the phenomenon of chromosome elimination. IV. The transfer of cytoplasmic genetic information from one to a second line or species. The latter aspect is already discussed by Galun (Chapter 10) and therefore this article will concentrate only on the first three problems.

II. SELECTION OF SOMATIC HYBRIDS

Induced fusions of protoplasts from two genetically different lines or species must necessarily result in a variety of homo- as well as heterokaryotic fusion products. The selection

of the few true somatic hybrid colonies from the mixed population of regenerating protoplasts is a key step in successful
somatic hybridization experiments. Two principal selection procedures for the isolation of somatic hybrids have been established (Table 1). In the first, selection of the somatic hybrids depends on the appearence of a different phenotype or
growth pattern in comparison to the parental cell colonies or
plantlets developing simultaneously (complementation selection).
In the second, direct mechanical selection of the fusion products. Both methods aspire to isolate the somatic hybrids as
early as possible.

A. *The Use of Mutants*

     The fusion of protoplasts of two nonallelic chlorophyll de-
ficient or auxotrophic mutants leads to somatic hybrids expressing the wild type phenotype (Table 2,3). The first experiments
to select somatic hybrids with the aid of albino mutants were
performed on haploid tobacco by Melchers and Labib (1974). Several intra- and interspecific somatic hybrids have been reported in *Nicotiana, Petunia* and *Datura* using nonallelic albino
mutants (Table 2). The use of auxotrophic mutants for the selection of somatic hybrids is restricted to three intraspecific
combinations (Table 3), though the use of such mutants is more
efficient because only the hybrid lines survive. Unfortunately,
it is difficult to obtain such mutants in higher plants and is
limited only in few plant species (see Maliga, Chapter 11).
Resistant mutants against chemicals such as antibiotics or amino acid analogues are more common. Successful somatic hybridization by the use of two different resistant mutants leading
to a double resistant somatic hybrid has been reported only
in *Nicotiana sylvestris* (White and Vasil, 1979) and in *Daucus
carota* (Harms *et al.*, 1981; Table 4).
     It is not always necessary to fuse two nonallelic mutants
for the selection of somatic hybrids. Cocking *et al.* (1977) fused protoplasts of an albino mutant of *Petunia hybrida,* which
can be regenerated in defined medium to shoots, with wild-type
protoplasts of *Petunia parodii,* which under the chosen culture
condition develop only into cell aggregates of at best 50 cells.
The selection of the somatic hybrids was based on their ability
to grow and on their green color. A similar selection system
was also employed for getting interspecific somatic hybrids of
several other *Petunia* species, and in *Nicotiana* and *Datura*
(Table 5). The same selection procedure was employed for intergeneric somatic hybridization experiments between carrot and
Aegopodium podagraria and *Petroselinum hortense,* respectively,
(Dudits *et al.* 1979 and 1980). In several combinations somatic

TABLE 1: *Principal Selection Systems for Somatic Hybrids*

I. *Complementation*
   a. *Auxotroph Ab + auxotroph aB* --- *wild type AaBb*
   b. *Albino Ab + Albino aB*   --- *wild type AaBb*
   c. *Resistant R_1r_2+ resist.r_1R_2* -- *double resist. $R_1r_1R_2r_2$*
   d. *Phenotype A + phenotype B* --- *phenotype C*
   e. *No regeneration + no regen.* --- *regeneration*

II. *Culture of Isolated Hybrid Protoplasts*

---

   a. *Single culture of hybrid protoplasts*
   b. *Nurse culture*

TABLE 2: *Somatic Hybrids Selected by the Use of Two Nonallelic Albino Mutants*

Albino   +   Albino   ------   Wild type

Species fused	Regeneration	References	
Nicotiana tabacum +			
Nicotiana tabacum	*plants*	*Melchers and Labib*	*(1974)*
		Gleba et al.	*(1975)*
Nicotiana tabacum +			
Nicotiana sylvestris	*plants*	*Melchers*	*(1977a)*
Nicotiana tabacum +			
Nicotiana rustica	*plants*	*Douglas et al.*	*(1981a)*
Nicotiana tabacum +			
Nicotiana glauca	*plants*	*Evans et al.*	*(1980)*
Petunia parodii +			
Petunia hybrida	*plants*	*Cocking*	*(1978)*
Datura innoxia +			
Datura innoxia	*plants*	*Schieder*	*(1977)*

---

hybrids could be obtained even in such cases where protoplasts
of the wild type species showed no divisions at all under the
chosen culture conditions, as in hybridization experiments between *Datura innoxia* and three other *Datura* species.

Protoplasts of albino mutants can also be fused with green
wild type protoplasts which can regenerate in culture. In such
cases selection of the somatic hybrids is based on their green
color and their different phenotype in comparison to that of
the regenerants developed simultaneously from the protoplasts
of the wild type partner (Table 6). For example, protoplasts
from an albino mutant of *Datura innoxia* were fused with green
wild type protoplasts of *Atropa belladonna* since no albino
mutant of *A. belladonna* was available. The selection of somatic
hybrid calli was based on their green color coded by the genome
of *A. belladonna* and the appearance of hairs coded by the genome of *D. innoxia* (Krumbiegel and Schieder, 1979). In all
other combinations (Table 6), however, selection of the somatic
hybrids was undertaken after the development of shoots.

Resistant factors also can be used in combination with wild
type protoplasts for the selection of somatic hybrids (Table 7).
As already mentioned, the protoplasts of *Petunia parodii* do not
form cell aggregates of more than 50 cells in defined medium.
This limited development is not suppressed by the addition of
1mg/l actinomycin D. The protoplasts of *P. hybrida,* which normally can be regenerated to plants, do not divide in the presence of this antibiotic. After fusion of the *P. parodii* and
P. hybrida protoplasts, several actinomycin D-resistant calli
could be selected which were regenerated to plants (Power *et
al.,* 1976).

TABLE *3: Somatic Hybrids Selected by the Use of Two
Nonallelic Auxotrophic Mutants*

Auxotroph + Auxotroph ------ Autotroph

Species fused	*Regeneration*	*References*
Sph. donnellii + Sphaerocarpos donnellii	plants	*Schieder* (1974)
Physcomitrella patens + Physcomitrella patens	plants	*Grimsley et al.(1977a,b)*
Nicotiana tabacum + Nicotiana tabacum	plants	*Glimelius et al.(1978)* *Wallin et al.(1979)*

TABLE 4: *Somatic Hybrids Selected by the Use of Two Different Mutants Resistant against Toxic Substances*

Resistant	+	Resistant	-----	Double resistant

Species fused	Regeneration	References
Nicotiana sylvestris + Nicotiana sylvestris	callus	White and Vasil (1979)
Daucus carota + Daucus carota	callus	Harms et al. (1981)

TABLE 5: *Somatic Hybrids Selected by the Use of Protoplasts from an Albino Mutant and Wild-Type Protoplasts (R=Protoplasts are able to regenerate; r=Protoplasts can not regenerate)*

Albino R	+	Wild type r	------	Wild type Rr

Species fused	Regeneration	References
Petunia hybrida + Petunia parodii	plants	Cocking et al. (1977)
Petunia inflata + Petunia parodii	plants	Power et al. (1979)
Petunia parviflora + Petunia parodii	plants	Power et al. (1980)
Nicotiana tabacum + Nicotiana knightiana	plants	Maliga et al. (1978)
Datura innoxia + Datura stramonium	plants	Schieder (1978a)
Datura innoxia + Datura discolor	plants	Schieder (1978a)
Datura innoxia + Datura sanguinea	shoots	Schieder (1980a)
Datura innoxia + Datura candida	plants	Schieder (1980a)
Daucus carota + Aegopodium podagraria	plantlets	Dudits et al. (1979)
Daucus carota + Petroselinum hortense	plantlets	Dudits et al. (1980)

B. The Use of Wild-Type Protoplasts

Somatic hybridization by the use of the wild-type protoplasts of *Nicotiana glauca* and *N. langsdorffii* was reported by
Carlson *et al.* (1972). The selection of the somatic hybrids was
based on the ability of the hybrid colonies to develop on hormone-free media in contrast to the parental colonies which
grew only on hormone-supplemented medium. Smith *et al.* (1976)
confirmed these results, but with modified techniques and the
somatic hybrids showed a higher chromosome number in comparison
to the report of Carlson *et al.* (1972).

Similar selection procedures have been undertaken in interspecific hybridization experiments by Power *et al.* (1977) on
Petunia and by Butenko and Kuchko (1980) on *Solanum* (Table 8).
Such systems, in which the different growth patterns of the parents and the hybrids in defined media are employed for the selection of somatic hybrids, are useful only in those instances
where sexual hybrids are available. Knowledge of the growth
pattern of the hybrids in defined media is necessary for selection. Obviously, the use of such a system is not possible between species that are sexually incompatible.

Some somatic hybrid cell lines and plantlets have also been
selected because of their abnormal morphology (Table 9). Binding and Nehls (1978), for example, selected somatic hybrid
colonies between *Vicia faba* and *Petunia hybrida* showing an intermediate phenotype. However, one should not expect that in
all combinations a different phenotype of the hybrids in the
callus stage will be expressed. In several somatic hybridization experiments where the hybrids were selected by the use of
mutants, a vigorous growth pattern of the somatic hybrids could
be observed at the callus stage (Schieder, 1978 and 1980; Douglas *et al.*, 1981). The use of this phenomenon for the selection
of the somatic hybrids has been recommended (Schieder and Vasil, 1980) and has been successfully demonstrated after fusion
of the wild type protoplasts of *Datura innoxia* and *D. quercifo-
lia*. Amongst ca. 6,000 developed calli 5 showed a larger size
which were selected and regenerated to shoots. One of these
presumed somatic hybrids proved to be indeed a true somatic hybrid (Table 10, Schieder, unpublished).

C. Isolation by Visual Means

Kao (1977) introduced a successful method of mechanically
isolating heterokaryocytes and cultivating them individually
(Table 11). Fusion of colorless protoplasts of *Glycine max*, derived from a cell culture, with the green mesophyll protoplasts
of *Nicotiana glauca* facilitated the isolation of heterokaryotic
fusion products. The transfer of the fusion bodies with micro-

TABLE 6: *Somatic Hybrids Selected by the Use of Protoplasts from Albino Mutant and Wild-Type Protoplasts. The Selection of the Somatic Hybrids was Based on Their Green and Intermediate Phenotype*

Albino   +   Wild type   -----   Green and Intermediate Phenotype

Species fused	Regeneration	References
Daucus carota +		
Daucus capillifolius	plants	Dudits et al. (1977)
Lycopersicon esculentum +		
Solanum tuberosum	plants	Melchers et al. (1978)
Datura innoxia +		
Atropa belladonna	shoots	Krumbiegel and Schieder (1979)
Nicotiana tabacum +		
Nicotiana rustica	plants	Nagao (1978)
		Iwai et al. (1980)
Nicotiana tabacum +		
Nicotiana glutinosa	plants	Nagao (1979)
Nicotiana tabacum +		
Nicotiana alata	plants	Nagao (1979)

TABLE 7: *Somatic Hybrids Selected by the Use of Protoplasts from a Resistant Mutant and Wild-Type Protoplasts (R=Protoplasts are able to regenerate; r=Protoplasts can not regenerate)*

Wild type r   +   Resistant r   ------   Resistant Rr

Species fused	Regeneration	References
Petunia hybrida +		
Petunia parodii	plants	Power et al. (1976)
Nicotiana knightiana +		
Nicotiana sylvestris	plants	Maliga et al. (1977)
Nicotiana tabacum (tumor) +		
Nicotiana tabacum	shoots	Wullems et al.(1979)

TABLE 8: *Somatic Hybrids Selected by the Use of Selective Media*

| Wild type | + | Wild type | ----- | Wild type on selective media |

Species fused	Regeneration	References	
Nicotiana glauca + *Nicotiana langsdorffii*	plants	Carlson *et al.*	*(1972)*
		Smith *et al.*	*(1976)*
		Chupeau *et al.*	*(1978)*
Petunia hybrida + *Petunia parodii*	plants	Power *et al.*	*(1977)*
Solanum tuberosum + *Solanum chacoense*	plants	Butenko and Kuchko	*(1980)*

TABLE 9: *Somatic Hybrids Selected by Their Intermediate Phenotype*

| Wild type | + | Wild type | ----- | Intermediate Phenotype |

Species fused	Regeneration	References
Petunia hybrida + *Parthenocissus tri- cuspidata*	callus	Power *et al.* *(1975)*
Vicia faba + *Petunia hybrida*	callus	Binding and Nehls *(1978)*
Atropa belladonna + *Petunia hybrida*	shoots	Gosch and Reinert *(1978)*

pipets into Cuprac dishes containing many individual small wells allowed the culture and observation of individual cells. With these manipulations 20 hybrid cell lines were obtained. Similar results were obtained by the isolation of heterokaryons of *Arabidopsis thaliana* and *Brassica campestris* with micropipets (Gleba and Hoffmann, 1978). An alternative method to the single cultivation of heterokaryons is their "nurse culture" in sus-

TABLE 10: *Somatic Hybrids Selected by Their Vigorous Growth Pattern*

Wild type   +   Wild type   -----   Wild type with *Vigorous Growth*

Species fused	Regeneration	References	
Datura innoxia + *Datura quercifolia*	shoots	Schieder	*(1981)*

TABLE 11: *Somatic Hybrids Obtained by Visual Identification and Mechanical Isolation of the Fusion Bodies*

Wild type   +   Wild type   ----   Hybrid Protoplasts Isolated *Mechanically*

Species fused	Regeneration	References	
Glycina max + *Nicotiana glauca*	callus	Kao	*(1977)*
Arabidopsis thaliana + *Brassica campestris*	plantlets	Gleba and Hoffmann	*(1978)*
Nicotiana knightiana + *Nicotiana sylvestris*	plants	Menczel et al.	*(1978)*

pensions of phenotypically different cells. Such a system is independent of the complex media that are necessary for culture of single protoplasts (Kao, 1977; Gleba, 1978). Menczel *et al.* (1978) fused protoplasts of an albino and kanamycin-resistant cell line of *Nicotiana sylvestris* with green wild-type protoplasts of *N. knightiana*. Interspecific fusion products were transferred to protoplast cultures of the chlorophyll-deficient cell line of *N. sylvestris*. All green and shoot producing calli could be selected as somatic hybrids.

III. CONFIRMATION OF THE SOMATIC HYBRID NATURE

    The somatic hybrid nature of the selected regenerants can
be confirmed by several methods. In cases where sexual hybrids
are available a comparison with them could be undertaken (Power *et al.*, 1976; Carlson *et al.*, 1972). Often the intermediate
phenotype of flowers, leaves, trichomes, roots or tubers confirmed their hybrid nature (Smith *et al.*, 1976; Schieder, 1978b;
Power *et al.*, 1980; Dudits *et al.*, 1977; Melchers *et al.*, 1978).
Additional evidence for the hybrid nature of most of the selected somatic hybrids came from their higher chromosome numbers.
In general, diploid protoplasts have been used for the production of somatic hybrids in higher plants. Therefore, predominantly tetraploid somatic hybrids should be expected. Indeed
the double-diploid or amphidiploid chromosome set in intra- as
well as interspecific somatic hybrid cell lines or plants could
be found in *Nicotiana* (Melchers and Labib, 1974; Melchers,
1977b; Gleba *et al.*, 1975; Chupeau *et al.*, 1978; Carlson *et al.*,
1972), *Petunia* (Power *et al.*, 1977; Cocking *et al.*, 1977),
Daucus (Dudits *et al.*, 1977), *Datura* (Schieder, 1977, 1978a,
1980a), and *Solanum* (Butenko and Kuchko, 1980), together with
somatic hybrid lines that showed either higher ploidy levels or
aneuploid chromosome numbers. In the intergeneric combinations
Nicotiana glauca + *Glycine max, Arabidopsis thaliana* + *Brassica
campestris, Datura innoxia* + *Atropa belladonna,* and *Petunia hybrida* + *Vicia faba* the somatic hybrid nature of the selected
lines could be additionally confirmed by the different size of
the chromosomes of both parents (Kao, 1977; Gleba and Hoffmann,
1978; Krumbiegel and Schieder, 1979; Binding and Nehls, 1978).
    The different isoenzyme pattern after gel electrophoresis
of the parental species often was used for the confirmation of
the somatic hybrid nature of the selected regenerants: alcohol
dehydrogenases and aspartate aminotransferases(Wetter, 1977),
esterases and glucose-6-phosphate dehydrogenases (Maliga *et al.*,
1977), peroxidases (Dudits *et al.*, 1977), superoxid dismutases
(Douglas *et al.*, 1981b), and amylases (Lönnendonker and Schieder, 1980). The somatic hybrids between *Solanum tuberosum* and
Lycopersicon esculentum could be confirmed by the analysis of
fraction-1 protein (Melchers *et al.*, 1978).
    In few cases the hybrid nature of the selected somatic hy-
brids was demonstrated genetically. The intraspecific somatic
hybrids of tobacco that have been produced after fusion of haploid protoplasts of two genetically different albino mutants
(Melchers and Labib, 1974) showed, after self-crossing, an offspring in green and mutated seedlings comparable to that of the
diploid sexual hybrids (Melchers, 1977). The hybrid nature of
the tetraploid intraspecific somatic hybrids of two albino mutants of *Datura innoxia* was demonstrated by the use of anther

culture (Schieder, 1976b). Power *et al.* (1978) compared the flower color segregation after self crosses of sexual tetraploid hybrids of *Petunia hybrida* and *Petunia parodii* with that of two somatic hybrids. Minor differences were detected. In the progeny of the amphidiploid somatic hybrids of *Datura innoxia* and *D. stramonium* or *Datura discolor*, respectively, a small portion of albino seedlings was detected (Schieder, 1980b). For the hybridization experiments one albino mutant of *D. innoxia* was employed. The appearance of albino seedlings may be the result of the formation of a very small percentage of quadrivalents in meiosis I. The seed derived green seedlings, however, were uniform in three sexual generations and therefore one can say, that these somatic hybrids are the only two examples up to now which are self-fertile and can be propagated as somatic hybrids via seeds.

IV. CHROMOSOME ELIMINATION

    In intergeneric fusion products often the chromosomes of both parents can be distinguished by their different size. In such cases chromosome elimination, if it takes place, can be studied. The somatic hybrid cell lines of *Nicotiana glauca* and *Glycine max* produced by Kao (1977), in which the chromosomes of both parents can be easily distinguished, were found to be unstable. The number of chromosomes of *N. glauca* decreased during culture. Similar observations were made on somatic hybrid cell lines between *Vicia faba* and *Petunia hybrida* (Binding and Nehls, 1978), where at least a stable line could be obtained which contained the whole chromosome set of *V. faba* and only 4 chromosomes of *P. hybrida* (Binding, pers. communication). The chromosome numbers of the somatic hybrid cell lines of *Arabidopsis thaliana* and *Brassica campestris* as well as of *Datura innoxia* and *Atropa belladonna* appeared to be stable over a long period of culture (Gleba and Hoffmann, 1978; Krumbiegel and Schieder, 1979). However, with the appearance of shoots loss of chromosomes of *B. campestris* or *A. belladonna* could be observed (Gleba and Hoffmann, 1979; Krumbiegel and Schieder, 1981). This, however, could not be confirmed for all regenerated shoots of the somatic hybrids between *A. thaliana* and *B. campestris*. For the selection of the *D. innoxia* + *A. belladonna* somatic hybrid cell lines a nuclear-coded albino mutant of *D. innoxia* was employed. With the appearance of shoots albino sectors and totally albino shoots could be observed indicating the loss of chromosomes of *A. belladonna* (Krumbiegel and Schieder, 1981). However, one green line was obtained possessing only 6, and a second possessing only 4 chromosomes, of *A. belladonna* together with the whole chromosome set of *D. innoxia* which were totally stable for more than 10 months. Such results indicate that via intergeneric protoplast fusion some chromosomes can be transferred from one species to another.

The phenomenon of chromosome elimination in intergeneric fusion products seems to be useful not only for the transfer of some chromosomes but also for the transfer of some genetic information from one plant to another perhaps via somatic recombination. Dudits et al. (1979) fused protoplasts of a nuclear coded albino mutant of Daucus carota with green wild type protoplasts of Aegopodium podagraria which had never undergone divisions. Their selected green plantlets possessed the chromosomes of only Daucus carota. Several chemical studies proved that some genetic information of A. podagraria was present in the selected plants. Similar results were obtained after fusion of protoplasts from an albino mutant of Daucus carota with X-ray irradiated wild type protoplasts of Petroselinum hortense (Dudits et al., 1980). These encouraging results, summarized in Table 12, require further confirmation.

TABLE 12: Somatic Hybrids which Contain only a Part of the Genetic Information from the Second Parent

Species fused	Regeneration	References	
Petunia hybrida + Parthenocissus tri-cuspidata	callus	Power et al.	(1975)
Daucus carota + Aegopodium podagraria	plantlets	Dudits et al.	(1979)
Daucus carota + Petroselinum hortense	plantlets	Dudits et al.	(1980)
Arabidopsis thaliana + Brassica campestris	plantlets	Gleba and Hoffmann	(1979)
Vicia faba + Petunia hybrida	callus	Binding and Nehls	(1978)
Datura innoxia + Atropa belladonna	plantlets	Schieder and Krumbiegel	(1981)

V. CONCLUSIONS

The results of somatic hybridization experiments presented here are limited only to model plants predominantly from the genera *Nicotiana, Petunia, Daucus, Solanum* and *Datura*. Somatic hybridization experiments have not yet been done with important crop plants. This is because of the limited number of species from which protoplasts can be regenerated to plants. However, the results obtained up to now demonstrate that it is possible to recover fertile and stable amphidiploid somatic hybrids after protoplast fusion (e.g. *Datura innoxia* + *D. stramonium* and *D. discolor*, respectively). That intraspecific somatic hybridization can be used for plant improvement vegetatively propagated species, is obvious. For example, Wenzel *et al.* (1979) proposed a breeding scheme for potato that combines classical breeding methods with parthenogenetic and androgenetic reduction of chromosomes, followed by somatic hybridization. The breeding scheme is based on the addition, by protoplast fusion, of different genomes containing various resistant genes.

The results obtained in intergeneric fusion experiments leading to asymmetric hybrids after partial or total chromosome elimination of one fusion partner opens the possibility of incorporating via protoplast fusion only parts of the nuclear genetic information from one species into another. However, for further successful somatic hybridization in the direction of plant improvement, it is essential to increase the number of species from which protoplast regeneration into plants is possible. This is especially necessary for the important crop plants, such as the cereals (see Vasil, Chapter 9) and the legumes.

REFERENCES

Binding, H., and Nehls, R. (1978) *Molec. Gen. Genet. 164,* 137.
Butenko, R.G., and Kuchko, A.A. (1980). *In* "Advances in Protoplast Research" (L. Ferency, G.L. Farkas, eds.), p. 293, Pergamon Press, Oxford.
Carlson, P.S., Smith, H.H., Dearing, R.D. (1972). *Proc. Natl. Acad. Sci.* U.S.A. *69,* 2292.
Chupeau, Y., Missonier, C., Hommel, M.-C., and Goujaud, J. (1978). *Molec. Gen. Genet. 165,* 239.
Cocking, E.C. (1978). *In* "Frontiers of Plant Tissue Culture" (T.A. Thorpe, ed.), p. 151, Univ. of Calgary Press, Calgary.
Cocking, E.C., George, D., Price-Jones, M.J., and Power, J.B. (1977). *Plant Sci. Lett. 10,* 7.
Douglas, G.C., Keller, W.A., and Setterfield, G. (1981a). *Can.*

J. Bot. 59, 220.

Douglas, G.C., Wetter, L.R., Nakamura, C., Keller, W.A., and Setterfield, G. (1981b). *Can. J. Bot. 59*, 228.

Dudits, D., Hadlaczky, G., Lévi, E., Fejér, O., Haydu, Z., and Lázár, G. (1977). *Theor. Appl. Genet. 51*, 127.

Dudits, D., Hadlaczky, G., Koncz, C., Lázár, G., and Horváth, G. (1979). *Plant Sci. Lett. 15*, 101.

Dudits, D., Fejér, O., Hadlaczky, G., Koncz, C., Lázár, G., and Horváth, G. (1980). *Molec. Gen. Genet. 179*, 283.

Evans, D.A., Wetter, L.R., and Gamborg, O.L. (1980). *Physiol. Plant. 48*, 225.

Gleba, Y.Y. (1978). *Naturwissenschaften 65*, 158.

Gleba, Y.Y., and Hoffmann, F. (1978). *Molec. Gen. Genet. 165*, 257.

Gleba, Y.Y., and Hoffmann, F. (1979). *Naturwissenschaften 66*, 547.

Gleba, Y.Y., Butenko, R.G., and Sytnik, K.M. (1975). *Dokl. Akad. Nauk. SSSR. 221*, 1196.

Glimelius, K., Eriksson, T., Grafe, R., and Müller, A.J. (1978). *Physiol. Plant. 44*, 273.

Gosch, G., and Reinert, J. (1976). *Naturwissenschaften 63*, 534.

Grimsley, N.H., Ashton, N.W., and Cove, D.H. (1977a). *Molec. Gen. Genet. 154*, 97.

Grimsley, N.H., Ashton, N.W., and Cove, D.H. (1977b). *Molec. Gen. Genet. 155*, 103.

Harms, C.T., Potrykus, I., and Widholm, J.M. (1981). *Z. Pflanzenphysiol. 101*, 377.

Iwai, S., Nagao, T., Nagata, K., Kawashima, N., and Matsuyama, S. (1980). *Planta 147*, 414.

Kao, K.N. (1977). *Molec. Gen. Genet. 150*, 225.

Kao, K.N., and Michayluk, M.R. (1974). *Planta 115*, 355.

Keller, W.A., and Melchers, G. (1973). *Z. Naturforsch. 28C*, 737.

Krumbiegel, G., and Schieder, O. (1979). *Planta 145*, 371.

Krumbiegel, G., and Schieder, O. (1981). *Planta, in press*.

Lönnendonker, N., and Schieder, O. (1980). *Plant. Sci. Lett. 17*, 135.

Maliga, P., Lázár, G., Joó, F., Nagy, A.H., and Menczel, L. (1977). *Molec. Gen. Genet. 157*, 291.

Maliga, P., Kiss, Z.R., Nagy, A.H., and Lázár, G. (1978). *Molec. Gen. Genet. 163*, 145.

Melchers, G. (1977a). *In* "International Cell Biology" (B.R. Brinkley, K.R. Porter, eds.), p. 207, Rockefeller Univ. Press, New York.

Melchers, G. (1977b). *Naturwissenschaften 64*, 184.

Melchers, G., and Labib, G. (1974). *Molec. Gen. Genet. 135*, 277.

Melchers, G., Sacristan, M.D., and Holder, A.A. (1978). *Carlsberg Res. Commun. 43*, 203.

Menczel, L., Lázár, G., and Maliga, P. (1978). *Planta 143*, 29.

Nagao, T. (1978). *Jap. J. Crop Sci. 48*, 385.

Nagao, T. (1979). *Jap. J. Crop Sci. 47,* 491.
Power, J.B., Frearson, E.M., Hayward, C., and Cocking, E.C. (1975). *Plant Sci. Lett. 5,* 179.
Power, J.B., Frearson, E.M., Hayward, C., George, D., Evans, P.K., Berry, S.F., and Cocking, E.C. (1976). *Nature 263,* 500.
Power, J.B., Berry, S.F., Frearson, E.M., and Cocking, E.C. (1977). *Plant Sci. Lett. 10,* 1.
Power, J.B., Sink, K.C., Berry, S.F., Burns, S.F., and Cocking, E.C. (1978). *J. Hered. 69,* 393.
Power, J.B., Berry, S.F., Chapman, J.V., Sink, K.C., and Cocking, E.C. (1979). *Theor. Appl. Genet. 55,* 97.
Power, J.B., Berry, S.F., Chapman, J.V., and Cocking, E.C. (1980). *Theor. Appl. Genet. 57,* 1.
Schieder, O. (1974). *Z. Pflanzenphysiol. 74,* 357.
Schieder, O. (1977). *Planta, 137,* 253.
Schieder, O. (1978a). *Molec. Gen. Genet. 162,* 113.
Schieder, O. (1978b). *Planta 141,* 333.
Schieder, O. (1980a). *Z. Pflanzenphysiol. 98,* 119.
Schieder, O. (1980b). *Molec. Gen. Genet. 179,* 387.
Schieder, O. (1981). Unpublished.
Schieder, O., and Vasil, I. (1980). *In* "Perspective in Plant Cell and Tissue Culture" (I. Vasil, ed.), p. 21, Intern. Rev. Cytol. 11B, Academic Press, New York.
Smith, H.H., Kao, K.N., and Combatti, N.C. (1976). *J. Hered. 67,* 123.
Vasil, I., and Vasil, V. (1980). *In* "Perspective in Plant Cell and Tissue Culture" (I. Vasil, ed.), p. 1, Intern. Rev. Cytol. 11B, Academic Press, New York.
Wallin, A., Glimelius, K., and Eriksson, T. (1979). *Z. Pflanzenphysiol. 91,* 89.
Wenzel, G., Schieder, O., Przewoźny, T., Sopory, S.K., and Melchers, G. (1979). *Theor. Appl. Genet. 55,* 49.
Wetter, L.R. (1977). *Molec. Gen. Genet. 150,* 231.
White, D.W.R., and Vasil, I. (1979). *Theor. Appl. Genet. 54,* 239.
Wullems, G.J., Molendijk, L., and Schilperoort, R.A. (1979). *Theor. Appl. Genet. 56,* 203.

CHAPTER 13

PLANT CELL TRANSFORMATIONS AND GENETIC ENGINEERING

Jeff Schell[1]
Marc Van Montagu
Marcelle Holsters
Ann Depicker
Patricia Zambryski
Patrick Dhaese

Laboratory of Genetics,
Rijksuniversiteit Gent,
Gent, Belgium

Jean-Pierre Hernalsteens
Jan Leemans
Henri De Greve

Laboratory GEVI,
Vrije Universiteit Brussel,
St.-Genesius-Rode, Belgium

Lothar Willmitzer
Leon Otten
Jo and Gudrun Schröder

Max-Planck-Institut für Züchtungsforschung,
Köln, FRG

I. INTRODUCTION

The genetic transformation of plant cells by experimental procedures is a basic step in the broader context of plant genetic engineering. The following general approaches

[1] Present address: Max-Planck-Institut für Züchtungsforschung, Köln, FRG.

can and have been used to introduce genes in eukaryotic cells.

A.   Direct uptake of DNA from solution or after coprecipitation with calcium phosphate

Although there is a fairly large body of literature concerning the use of these methods with plant cells and protoplasts, most, if not all of this work is inconclusive because the transforming DNA was either not sufficiently characterized and/or because a careful and critical analysis of the transformed cell lines by, e.g., Southern gel blotting DNA/DNA hybridizations, was not presented.

B.   DNA viruses or double-stranded cDNA copies of RNA viruses as host gene vectors

Only very few plant DNA viruses are known. The best studied virus is CaMV, which infects some crucifer plants. Its genome consists of a small (± 8 Kb) double-stranded circular DNA. The entire genome of CaMV was cloned in an E. coli vector and this cloned genome was shown to be infective on turnips (Howell et al., 1980). The whole of the nucleotide sequence has been worked out by Franck et al. (1980) and work is in progress to try to develop this virus genome as a gene vector. For a recent review, see Hohn and Hohn (1982). To the best of our knowledge, no reports have as yet been published about the use of cDNA copies of RNA plant viruses as potential gene vectors. Some single-stranded DNA viruses of cereals (gemini viruses; see Goodman, 1981) should also be considered but no successful application as gene vectors has been published thus far.

C.   Isolated DNA sequences harboring genes that can be used as selectable markers to identify transformed cells

Several groups are currently attempting to develop such marker genes. One promising model is the nitrate reductase (NR) gene. Nitrate reductase-deficient cell lines have been isolated (e.g. in tobacco) and selective conditions have been worked out to select for cells in which NR activity has been restored (Mendel and Müller, 1979). Such a cloned NR gene has, however, not yet been

isolated. Recently, one has started to look for drug resistance markers as selectable marker genes.   The prominent
candidates are bacterial genes coding for phosphorylation
or acetylation enzymes that inactivate antibiotics as
kanamycin, neomycin, G418 and gentamycin.   The fact that
some of these genes have recently been found to be expressed in yeast cells and some others in mammalian cells, lends
support to efforts to use these or similar marker genes for
plant cells.   No success has, however, been reported yet
and we do not know whether plant cells will readily take up
DNA of isolated genes without significant degradation and,
if so, whether the bacterial genes will be readily expressed.   The use of liposomes to introduce DNA in plants has
recently been discussed (Lurquin, 1981).

D.   Micro-injection

The direct injection of DNA in the nuclei of animal
cells has been developed over the last years into an extra-
ordinary potent tool.   No successful attempt to use this
method for plant cells has yet been described.

E. The Ti plasmids of A. tumefaciens

For dicotyledonous plants this natural gene vector has
been successfully used to experimentally introduce isolated
genes into the chromosomes of plant cells.   A review of the
properties of Ti plasmids as gene vectors will therefore be
presented in this paper.

II.   THE Ti-PLASMID AS A NATURAL GENE VECTOR

Ti plasmids are harbored by a group of soil bacteria
(Agrobacteria) and are responsible for the capacity of
these bacteria to induce so-called "crown gall" tumors on
most dicotyledonous plants. (For recent reviews, see Gordon,
1979; Schilperoort et al., 1979; Schilperoort et al., 1980;
Schell et al., 1979; Van Montagu and Schell, 1979; Schell
and Van Montagu, 1980; Van Montagu et al., 1980). Crown
gall tumors proliferate autonomously in tissue culture on
simple media devoid of growth hormones.   The tumorous state
of crown gall tissue results from the transfer of Ti plasmid DNA from Agrobacterium tumefaciens to the plant cells.

This system represents a natural gene-vector for plant
cells, evolved by and for the benefit of the bacteria that
harbor Ti plasmids (Schell et al., 1979). Crown gall cells,
as a direct result of genetic transformation by the Ti plasmid, produce various substances called opines. Free-living
Agrobacteria utilize these opines as sources of carbon and
nitrogen. The genetic information both for the synthesis
of opines in transformed plant cells and for their catabolism by free-living Agrobacteria is carried by Ti plasmids. Because different Ti plasmids induce and catabolize
different opines, the Ti plasmids can be classified according to the type of opine they determine (Guyon et al.,
1980).

Bacteria are known to be able to conquer an ecological
niche by acquiring the capacity to catabolize certain organic compounds, not readily degradable by most other bacterial species. In several cases, the genes determining
this degradative capacity have been found to be part of
extrachromosomal plasmids. Several groups of soil bacteria,
especially those living in and around the rhizosphere of
plants, are able to decompose organic compounds released by
plants. Clearly, with the advent of Ti plasmids, Agrobacteria have carried this capacity one step further, by genetically forcing plant cells - via a gene transfer mechanism - to produce specific compounds (opines) which they are
uniquely equipped to catabolize. This novel type of parasi-
tism has therefore been called "genetic colonization"
(Schell et al., 1979).

Ti plasmids are responsible for the following properties of Agrobacteria : (i) crown gall tumor induction; (ii)
specificity of opine synthesis in transformed plant cells;
(iii) catabolism of specific opines; (iv) agrocin sensitivity; (v) conjugative transfer of Ti plasmids; and (vi)
catabolism of arginine and ornithine (Zaenen et al., 1974;
Van Larebeke et al., 1974, 1975; Schell, 1975; Watson et
al., 1975; Engler et al., 1975; Bomhoff et al., 1976;
Genetello et al., 1977; Kerr et al., 1977; Petit et al.,
1978a, 1978b; Firmin and Fenwick, 1978; Klapwijk et al.,
1978; Ellis et al., 1979; Guyon et al., 1980).

In order to identify the relevant plasmid genes,
mutant plasmids were isolated by transposon-insertion mutagenesis (Van Montagu and Schell, 1979; Hernalsteens et al.,
1978; Holsters et al., 1980; Ooms et al., 1980; Garfinkel
and Nester, 1980; De Greve et al., 1981) and by deletion
(Holsters et al., 1980; Koekman et al., 1979). The different mutations were localized on the physical maps established for nopaline (Depicker et al., 1980) and octopine
(De Vos et al., 1981) Ti plasmids, thus allowing the estab-

lishment of functional genetic maps for these plasmids (Holsters et al., 1980; De Greve et al., 1981). Two aspects of the transfer of the so-called T-region of Ti plasmids to the nucleus of plant cells were recently studied in greater detail : the integration of T-DNA in chromosomal plant DNA and the expression of T-DNA in transformed plant cells.

A.   The integration of the T-region of Ti plasmids in
      chromosomal plant DNA

The T-region is defined as that segment from the Ti plasmids that is homologous to sequences present in crown gall cells. The sequences which are transferred from the Ti plasmid to the plant and determine tumorous growth have been called T-DNA. The T-region of octopine and nopaline Ti plasmids has been studied in most detail both physically and functionally and Figure 1 summarizes these results. The T-regions, roughly 23 kb in size, are only a portion of the entire plasmids (Lemmers et al., 1980; Thomashow et al., 1980a; De Beuckeleer et al., 1981; Engler et al., 1981). Southern blotting and cross-hybridization of restriction endonuclease digests of the two types of plasmids as well as electron microscope heteroduplex anal-yses has revealed that 8 to 9 kb of the T-DNA regions are conserved (Chilton et al., 1978; Depicker et al., 1978; Engler et al., 1981); these sequences are represented as the shaded areas in the figure. So far, all attempts to reveal homology between the T-DNA region and plant DNA have failed.

The transfer of DNA from Agrobacterium to plant cell DNA includes at least two stages - a primary interaction between bacterial and plant cell walls presumably leads to the transfer of the whole or part of the Ti plasmid to the plant cell, and a secondary reaction must involve the insertion of the T-DNA of the Ti plasmid into the plant cell genome. The only possibility for observing this event is essentially to compare the T-region of the Ti plasmid in bacteria with the final T-DNA structure after it has ar-rived in the plant.

Although only 4 nopaline tumor lines have been anal-yzed in detail (Lemmers et al., 1980; Zambryski et al., 1980, and submitted; Yadav et al., 1980), the data suggest that the mechanism of T-DNA transfer is rather precise since the same continuous segment of the Ti plasmid is always present. Some nopaline tumor lines appear to con-tain a single T-DNA copy. In two other nopaline tumor lines the T-DNA occurs in multiple copies which are organ-

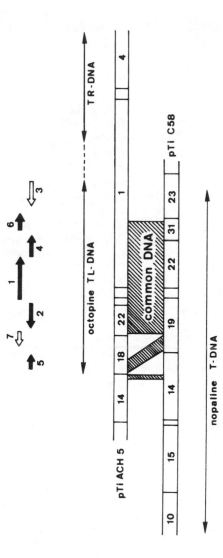

FIGURE 1. The T-regions of octopine and nopaline Ti plasmids have been aligned; shaded areas indicate the DNA sequences that are common to both T-regions (Engler et al., 1981). The numbers refer to fragments produced following digestion with HindIII. The arrows above the figure represent different transcripts of T-DNA in octopine crown gall tumors. Black arrows represent transcripts derived from common DNA sequences.

ized in a tandem array; this was demonstrated most clearly by molecular cloning of T-DNA fragments from transformed plant cell DNA which contain sequences derived from the right and left ends of the T-DNA region of the Ti plasmid. To date we have more detailed information on the boundaries of nopaline T-DNAs since the nucleotide sequence of these borders has been determined by comparison with the nucleotide sequence of the same region of the nopaline Ti plasmid (see below).   Sequence analysis of left-right junctional clones derived from the tandemly organized T-DNAs has revealed that there can be reorganization of the sequences contained at the ends of the T-DNA.

A single T-DNA copy was observed in a tumor line which was induced using a Ti plasmid containing the transposon Tn7 in the right end of the T-region resulting in the loss of nopaline synthase activity in the tumor cells (Lemmers et al., 1980; Hernalsteens et al., 1980; Holsters et al., 1980).   In this tumor line T-DNA is enlarged by the exact size of 15 kb of Tn7 and molecular cloning suggests that the borders of this T-DNA occur in the same region of Ti plasmid sequences as observed for wild-type nopaline Ti plasmids (Holsters et al., 1982).   The generation of tandem copies of T-DNA in some tumor lines and the integration of internally enlarged T-DNAs both with similar boundaries suggest that the ends of the T-DNA in nopaline tumor lines are well defined.

There have been a few more octopine tumor lines studied and the data suggest that the octopine T-DNA is more variable (Thomashow et al., 1980a, 1980b; De Beuckeleer et al., 1981; Ooms et al., submitted).   A left T-DNA region (TL) containing the conserved sequence (as in the nopaline T-DNA region) is always present; this region is usually 12 kb in size but one Petunia tumor line is shortened at the right end of TL by about 4 kb (De Beuckeleer et al., 1981).   In addition, there is often a right T-DNA region (TR) which contains sequences which are adjacent but not contiguous in the octopine Ti plasmid and in one tumor line TR is amplified (Merlo et al., 1980; Thomashow et al., 1980a).   Recent observations indicate that TL can also be a part of a tandem array (M. Holsters, unpublished results). The dotted horizontal line in the figure indicates that it is unknown whether TL and TR are linked together in the plant chromosome.   Both the interruption of Ti plasmid sequences in the generation of TR and TL, and the existence of an amplified TL and TR indicate that reorganization of octopine T-DNA sequences also occurs.

It is obvious that the T-DNAs of nopaline and octopine tumor lines are not identical.   However, there are some

similarities : both T-DNAs contain a common or conserved
DNA and both T-DNAs can be amplified either in whole as in
the nopaline T-DNA or in part as in octopine T-DNA. It is
unknown whether the integration is the result of plant or
Ti plasmid specific functions but it is not unlikely that
both are involved.

   We do not yet have sufficient information to make
elaborate models for T-DNA transfer. The sequences of the
left and right border regions of the nopaline Ti plasmid
have been determined (Zambryski et al., 1980, and submitted). The data from the exact nucleotide sequence of
4 right and 4 left T-DNA borders isolated from nopaline
tumor lines suggests that the mechanism of T-DNA integration and subsequent stabilization is precise on the right
and less precise on the left. The T-DNA/plant DNA junction
in the right has been shown to occur at exactly the same
base pair with respect to Ti plasmid sequences in the T-DNA
isolated from two independent tumor lines. Only a variation of a single base was found in the analysis of the
right border of a junctional region containing the left-
right T-DNA boundary from a T-DNA which was part of a
tandem array. All the left T-DNA boundaries we have determined so far are different. The results from the sequence
analysis do not reveal how T-DNA transfer has occurred;
nevertheless, the data suggest that well-defined regions of
the Ti plasmid are involved in this process.

   The nucleotide sequence of the T-region around the
borders of the nopaline Ti plasmid has also been determined
(Zambryski et al., submitted) (data not shown). It is easy
to find the region of interest on the right side of the
T-region since the T-DNA border in this side varies over a
single base pair. On the left we have found 3 borders
spanning a region of 6 base pairs. However, a T-DNA border
which occurred further to the left has been isolated and
sequenced (M.-D. Chilton, personal communication).

   One can compare the sequences within 100 base pairs of
either the right T-DNA border region or the left T-DNA
border region of the nopaline Ti plasmid. There is a short
direct repeat of 12 base pairs which can be extended to
22 base pairs with 3 mismatched bases which is found on
either side of the T-DNA region of the nopaline Ti plasmid.
The 22 base pairs have the following sequence :

```
                 G C   G
     C A G G A T A T A T T G T G G T G T A A C

     G T C C T A T A T A A C A C C A C A T T G
                 C G   C
```

This sequence is found just one base pair to the right of the T-DNA border on the right side of the nopaline Ti plasmid and 100 base pairs to the left of the left T-DNA border on the left side of the nopaline Ti plasmid. The left T-DNA border described by Chilton et al. (in preparation) in fact ends within these 22 base pairs sequence to exclude the first 12 base pairs. A very similar sequence has also been observed near the left border of the octopine T-region (unpublished data) Future studies of mutants where these sequences are specifically altered may establish whether these sequences are fundamental to T-DNA transfer.

Attention may also be due to the presence of a sequence just adjacent to the 22 base pair direct repeat at the left border of the T-DNA region of the nopaline Ti plasmid which is an exact Chi sequence (Stahl et al., 1980; Smith et al., 1981). Chi mainly stimulates recombination in prokaryotes and has the property of acting to influence the recombinogenic behaviour of DNA which is found at a distance from its own residence. Chi has even been proposed to be involved in the synthesis of hybrid immunoglobulins (Kanter and Birshtein, 1981). Whether or not this Chi sequence plays a role in T-DNA transfer has not been determined.

B.   Expression of T-DNA in transformed plant tissues

Previous work has shown that the T-DNA is transcribed in transformed plant cells (Drummond et al., 1977; Gurley et al., 1979; Willmitzer et al., 1981a; Gelvin et al., 1981), but the number, sizes, and location of the coding regions were not known. Recently, these questions were investigated in detail (Willmitzer et al., 1982). The cell suspension cultures from tobacco used in these experiments harbor the T-DNA from pTiA6, a Ti plasmid which induces octopine synthesis in transformed cells. The cells contain only the TL fragment of the T-region (Fig. 1). Tumor-specific RNAs were detected and mapped by hybridization of ^{32}P-labeled Ti plasmid fragments to polyadenylated RNA which had been separated on agarose gels and then transferred to DBM paper. The results, summarized in Fig. 2, show that the cells contain a total of seven distinct transcripts which differ in their relative abundance and in their sizes. They all bind to oligo(dT)-cellulose, indicating that they are polyadenylated. Thus, the T-DNA which was transferred from a prokaryotic organism must provide specific poly(A)-addition sites. The direction of transcription was determined for six of the seven transcripts,

Transcripts of T_L-DNA in octopine tumors

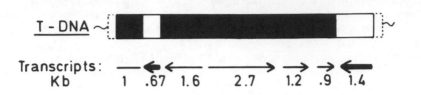

Fig. 2. Transcripts in octopine tumor line A6-S1.   The
numbers refer to the size of the RNAs in bases   x 10^{-3}.
The thickness of the lines indicates the relative abundance
of the RNAs, and the arrows describe the transcription
direction of the six transcripts where this has been determined.

and the location of the approximate 5'- and 3'-ends were
mapped on the TL-DNA.
     All seven RNAs mapped within the T-DNA sequence.
This, and the observation that transcription is inhibited
by low concentrations of α-amanitin (Willmitzer et al.,
1981b) seems to suggest that each transcript is determined
by a specific promoter site on the TL-DNA recognized by
plant RNA polymerase II.   The data presented here do not
rule out that some T-DNA promoters serve for transcription
of more than one RNA.   Considering the groupwise orienta-
tion of several transcripts (Fig. 2), the simplest model
would assume one promoter site per group of transcripts.
If so, one would expect that inactivation of a 5'-proximal
gene of a group would also lead to disappearance of the
transcripts from the 5'-distal genes.   However, analysis of
some cell lines containing the T-DNA of Ti plasmid mutants
indicates that groupwise inactivation of genes did not
occur (Leemans et al., 1982).   The results available so far
are consistent with the assumption that each gene of the
TL-DNA has its own signals for transcription in the eukaryotic plant cells.
     To understand the mechanisms of T-DNA action it is

important to know whether these RNAs are translated into proteins and to analyze their functions. The hybridization studies and a previous report on T-DNA-derived translatable mRNAs (McPherson et al., 1980) indicated that the concentrations of these RNAs are very low in transformed plant cells. It was therefore necessary to develop a hybridization selection procedure sufficiently sensitive and specific to detect mRNAs which represent about 0.0001% of the total mRNA activity in the plant cells. This procedure was used to enrich for T-DNA-derived mRNAs by hybridization to Ti plasmid fragments covalently bound to microcrystalline cellulose; the hybridized RNAs were eluted and translated in vitro in a cell-free system prepared from wheat germ.

The results obtained with this approach are summarized in Fig. 3 (Schröder and Schröder, 1982). The octopine tumor cells contained at least three T-DNA-derived mRNAs which are translated in vitro into distinct proteins, and the coding regions correlate with those of three transcripts. The protein encoded at the right end of the TL-DNA (Mr 39,000) was of specific interest since previous genetic analysis indicated that this part is responsible for octopine synthesis (Koekman et al., 1979; De Greve et al., 1981; Garfinkel et al., 1981) and since the size of the in vitro synthesized protein is identical with that of the octopine-synthesizing enzyme in octopine tumors. Immunological studies showed that this protein was recognized by antiserum against the tumor-specific enzyme (Schröder et al., 1981a). These results demonstrate that the structural gene for the octopine-synthesizing enzyme is on the Ti plasmid. So far, this is the only protein product of the T-DNA with known enzymatic properties; the possible functions of the two smaller T-DNA-derived proteins are not known. The region coding for the octopine-synthesizing enzyme has recently been sequenced (De Greve et al., in preparation). According to these data, the sequences for initiation and termination of transcription reveal signals of eukaryotic rather than prokaryotic type, suggesting that this gene is designed for expression in eukaryotic cells.

However, this is not necessarily true for all genes of the T-region, since transcripts were also detected in Agrobacteria (Gelvin et al., 1981). It was therefore of interest whether the T-DNA-derived mRNAs isolated from plant cells shared properties with typical prokaryotic or eukaryotic mRNAs. The fact that translation of all three mRNAs was inhibited by the cap analogue pm^{7}G suggests but does not prove that they contain a cap structure at the 5'-end. This would be typical for eukaryotic mRNA since caps have not been described in prokaryotic RNA. All three mRNAs

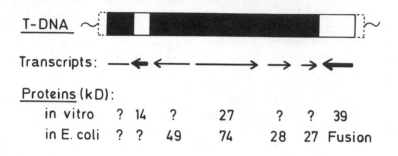

Fig. 3. Correlation of transcripts with proteins encoded on TL-DNA. Proteins were identified either by translation of hybrid-selected mRNA from tobacco cell line A6-S1 (in vitro) or by expression of coding regions in E. coli minicells (in E. coli). The numbers refer to the size of the proteins in Daltons x 10^{-3}. Question marks indicate that it has not been possible so far to demonstrate proteins correlating with the transcripts. The protein with Mr 39,000 represents the octopine-synthesizing enzyme.

were found in polyadenylated as well as in nonpolyadenylated RNA fractions; this cannot be used for such a tentative classification, since both types of RNA have been described in eukaryotic as well as in prokaryotic cells.

The mRNAs for the three proteins each represent about 0.0001% of the total mRNA activity in polyribosomal RNA, and this appears to be the detection limit at present for translatable RNA. The other four transcripts detected by hybridization experiments are present at even lower concentrations. This is likely to be the reason, assuming that they possess mRNA activity, why the corresponding proteins could not be identified by in vitro translation so far.

A different approach was therefore developed to search for coding regions on the T-DNA and their protein products. Fragments from the T-region were cloned into E. coli plasmids and analyzed for gene expression in E. coli minicells

(Schröder et al., 1981b). Some of the results are summarized in Fig. 3.   There are at least four different coding
regions within the TL-DNA which can be expressed in minicells into distinct proteins from promoters which are
active in the prokaryotic cells.   The four regions expressed in E. coli correlated with four regions transcribed
in plant cells into RNA.   The plant transcripts are larger
than the proteins in E. coli, and the regions expressed in
minicells appear to lie within the regions transcribed in
plant cells.   Although other explanations are not ruled
out, one may speculate that plant cells and E. coli express
at least partly the same coding regions.

III. THE DEVELOPMENT OF THE Ti PLASMID AS AN EXPERIMENTAL GENE VECTOR

A.   Genes inserted in the T-region of Ti plasmids are cotransferred to the plant nucleus as part of the T-DNA

In view of the observed involvement of the "ends" of
the T-region in the integration of T-DNA, it could be
expected that any DNA segment inserted between these "ends"
would be cotransferred, provided no function essential for
T-DNA transfer and stable maintenance was inactivated by
the insertion.   The genetic analysis of the T-region by
transposon-insertion provided Ti plasmid mutants to test
this hypothesis.   A Tn7 insertion in the nopaline synthase
locus, produced a Ti plasmid able to initiate T-DNA transfer and tumor formation (De Beuckeleer et al., 1978; Van
Montagu and Schell, 1979; Hernalsteens et al., 1980).
Analysis of the DNA extracted from these tumors showed
that the T-region containing the Tn7 sequence had been
transformed as a single 38 x 10^3 base pair segment without
any major rearrangements.   Several different DNA sequences
have since been introduced in different parts of the T-region of octopine and nopaline Ti plasmids (see e.g. Leemans
et al., 1981).   The preliminary observations with these
mutant Ti plasmids fully confirm the initial expectations :
most DNA sequences inserted between the "ends" of the
T-region are cotransferred with and become a stable part of
the T-DNA of the plant tumor cells transformed with these
mutant Ti plasmids.   If the experimental insert would
inactivate a function essential for the transfer of the
T-region or for the integration of the T-DNA, such a mutant

Ti plasmid would not be able to transform plant cells. No
such inserts have as yet been characterized, indicating
that T-DNA transfer and integration is either coded for by
genes outside of the T-region or by a very limited number
of genes in the T-region or by a combination of functions
with different genetic localization.

These observations have, therefore, firmly established
that the Ti plasmids can be used as experimental gene
vectors and that large DNA sequences (of at least 15 kb)
can be transferred stably to the nucleus of plant cells as
a single DNA segment.

B.   The T-DNA contains transcription promoter signals

As described under II.B., the T-DNA was found to be
transcribed in crown gall tissues.   The 5'- end of the
octopine synthase mRNA was accurately mapped by sequencing
of a T-region DNA fragment hybridized to this mRNA and thus
protected from degradation by the single strand specific S1
nuclease (De Greve et al., in preparation).   The complete
base sequence from the octopine synthase gene was determined (De Greve et al., in preparation).   The promoter
sequence that has thus been identified is more eukaryotic
than prokaryotic in its recognition signals and no introns
interrupt the open reading frame which starts at the first
AUG codon following the 5' start of the transcript. Similar
work was also performed for the nopaline synthase gene (M.
Bevan and M.-D. Chilton; L. Willmitzer; H. Goodman; P.
Zambryski; A. Depicker; S. Stachel; J. Gielen; P. Dhaese;
personal communications).   Essentially the same conclusions
were reached.

It is important to note that the promoter sequences
for the octopine synthase appear to escape possible control
mechanisms since they remain active in all tissues of
plants regenerated from tobacco cells transformed with
mutant Ti plasmids (De Greve et al., in preparation; see
also section III.C.). These observations are crucial to the
development of the Ti plasmid as a gene vector since they
provide us with the possibility to insert coding sequences
behind these promoters with a reasonable chance to see
these sequences expressed in plant cells.

In this respect we also studied the expression of the
Tn7 insertion in the nopaline synthase locus as a model
system for other genes inserted in this area. Tn7 codes for
a dihydrofolate reductase which is resistant to methotrexate (Tennhammer-Ekman and Sköld, 1979). Suspension
cultures of both untransformed and crown gall tobacco

tissue were found to be completely inhibited by 2 µg/ml methotrexate. In contrast, a culture established with crown gall cells containing Tn7 grew well on media containing 2 µg/ml methotrexate. In addition, nuclei isolated from these methotrexate resistant tobacco lines were shown to synthesize Tn7 transcripts homologous to Tn7 and a poly(A) mRNA fraction derived from purified polysomes was also shown to contain Tn7 transcripts.

C. The use of "intermediate vectors" for the in vitro introduction of selected genes in the T-region of Ti plasmids

Because of the size of Ti plasmids (about 130 Md) and because a large number of Ti plasmid genes are involved in the transformation mechanism, it did not appear to be feasible to develop a "mini-Ti" cloning vector with unique cloning sites at appropriate locations within the T-region and with all functions essential for T-DNA transfer and stable maintenance. An alternative way was therefore developed to introduce genes at specified sites in the T-region of a functional Ti plasmid. The principle was to make an "intermediate vector" (Van Montagu et al., 1980; Schell et al., 1981; Leemans et al., 1981) consisting of a common E. coli cloning vehicle, for example the pBR322 plasmid, into which an appropriate fragment of the T-region of a Ti plasmid is inserted. Single restriction sites in this T-region fragment can then be used to insert the chosen DNA sequence. This "intermediate vector" is subsequently introduced, via mobilization, into an A. tumefaciens strain carrying a Ti plasmid which has been made constitutive for transfer (Holsters et al., 1980) and which already carries antibiotic resistance markers (e.g. streptomycin, sulfonamide) cloned into the same T-region restriction site as that chosen to insert the DNA to be transferred. These resistance markers were introduced into this site by a procedure essentially identical to the one described here. Recombination in vivo will transfer the DNA of interest into the appropriate site of the Ti plasmid. The Ti plasmid, thus engineered to contain the desired DNA, will transfer this DNA into plant cells (Leemans et al., 1981). It should be noted that the introduction of inserts and substitutions in all parts of the T-region by these methods allowed us to probe for the function of the different T-DNA transcripts (Leemans et al., 1982).

D.   Idenfication and elimination of tumor-controlling genes

A detailed genetic analysis of the TL region of octopine crown galls (Leemans et al., 1982) when correlated
with an analysis of the number, size and map position of
TL-DNA derived, polyadenylated transcripts (Willmitzer et
al., 1982), has allowed us to assign functions to the seven
different, well-defined transcripts coded for by TL-DNA.

The two most abundant transcripts, i.e. transcripts 7
and 3 of respectively 670 and 1400 nucleotides, are specific for octopine tumors. Transcript 3 was found to code for
the enzyme LpDH which synthesizes octopine. The function of
transcript 7 is still unknown. The other five transcripts
are derived from TL-DNA sequences that are homologous with
an equivalent region in the T-DNA of nopaline tumors (see
Fig. 1).

Transcripts 4 and 6 of respectively 1200 and 900
nucleotides, have homologous counterparts of the same size
in nopaline tumors, and this is probably also the case for
transcripts 2 (1600 bp) and 5 (1000 bp). Transcript 1
(2700 bp) although coded for by an octopine TL-DNA sequence
which is largely homologous to an equivalent sequence in
nopaline T-DNA, could thus far not be detected in nopaline
tumors (Willmitzer et al., 1982).

By observing the properties of crown gall tumors
obtained by infection of tobacco with mutant octopine
Ti plasmids carrying deletions of specified segments of the
TL-region, it was possible to assign functions to each of
the transcripts 1, 2, 4, 6 and to suggest a possible function for transcript 5 (Leemans et al., 1982).

The main conclusion from these studies were :
1.   that most, if not all, of the transcripts were ex-
      pressed from individual promoters;
2.   that T-DNA transfer and tumorous growth are controlled
      by different and independently acting functions;
3.   that none of the TL-DNA derived transcripts are essen-
      tial for T-DNA transfer;
4.   that transcripts 1, 2, 4, 6 and possibly also 5, act
      by suppressing organ development;
5.   that shoot formation (transcripts 1 and 2) and root
      formation (transcript 4) are suppressed by the action
      of different transcripts;
6.   that transcripts 4 and 6 are sufficient to maintain
      the tumorous growth properties of transformed tobacco
      cells.

If these conclusions are correct, it follows that
elimination of the coding sequences for transcripts 1, 2,

4, 6 and 5 from the T-DNA of transformed cells should result in tobacco cells containing and expressing some of the T-DNA genes, e.g. transcript 3 (thus producing LpDH activity) but no longer suppressed for organ development. Such mutant TL-DNAs were recently obtained (De Greve et al., submitted; Schell et al., 1981). As predicted, they were shown to form fertile plants with normal organ development and expressing transcript 3 in most, if not all, of their cells.

IV. MENDELIAN INHERITANCE OF T-DNA LINKED GENES

A number of published observations (Braun and Wood, 1976; Yang et al., 1980) had been interpreted to indicate that T-DNA sequences in grafted teratomas were lost during meiosis in flowering grafts. This led to the important suggestion that T-DNA sequences might be excised during meiosis.

TABLE I. Transmission of octopine synthesis gene in tobacco plant rGV1 : the gene is transmitted to progeny as a dominant Mendelian gene

Crosses	No. of progeny tested	Octopine synthesis		
rGV1 ♂ x rGV1 ♀	145		110 (76 %)	35 (24 %)
rGV1 ♂ x rGV1 ♀	200	$++^a$ 42 (21 %)	$+a$ 95 (48 %)	- 63 (31 %)
rGV1 ♂ x wild-type ♀	248		124 (50 %)	124 (50 %)
rGV1 ♂ x wild-type ♀	187		81 (43 %)	106 (57 %)
Plantlets derived from anther cultures (haploid)	102		47 (46 %)	55 (54 %)

[a] as estimated from a semi-quantitative assay of enzyme activity

To test this possibility, sexual crossed were performed using flowering tobacco plant containing a partially deleted T-DNA and expressing transcript 3. It was found that the T-DNA was stably transmitted through meiosis, both through the pollen and through the eggs and segregated as a single dominant Mendelian gene (Otten et al., 1981) (Table I).

V. GENERAL CONCLUSIONS

With the crown gall system we are confronted with a well documented, natural instance of a translocation system involving DNA transfer between prokaryotes and eukaryotes. The T-DNA is a translocating element present in a prokaryotic plasmid but consisting of DNA sequences which become functional genes in eukaryotic cells.

It is interesting to speculate that other such natural instances of DNA translocations, involving widely different organisms, can be expected to occur in situations involving a close and long term symbiotic or parasitic association between these different organisms.

ACKNOWLEDGMENTS

The investigations reported here were supported by the Max-Planck Organization and by grants from the "Kankerfonds van de A.S.L.K", from the "Instituut tot aanmoediging van het Wetenschappelijk Onderzoek in Nijverheid and Landbouw" (248/A), from the "Fonds voor Wetenschappelijk Geneeskundig Onderzoek" (3.0052.78), and the "Onderling Overlegde Akties" (12052179). M.H. and J.P.H. are Research Associates of the Belgian National Fund for Scientific Research (N.F.W.O.).

REFERENCES

Bomhoff, G., Klapwijk, P.M., Kester, H.C.M., Schilperoort, R.A., Hernalsteens, J.P., and Schell, J. (1976). Molec. Gen. Genet. 145, 177.
Braun, A.C., and Wood, H.N. (1976). Proc. Natl. Acad. Sci. USA 73, 496.
Chilton, M.-D., Drummond, M.H., Merlo, D.J., and Sciaky, D. (1978). Nature 275, 147.

De Beuckeleer, M., De Block, M., De Greve, H., Depicker, A., De Vos, R., De Vos, G., De Wilde, M., Dhaese, P., Dobbelaere, M.R., Engler, G., Genetello, C., Hernalsteens, J.P., Holsters, M., Jacobs, A., Schell, J., Seurinck, J., Silva, B., Van Haute, E., Van Montagu, M., Van Vliet, F., Villarroel, R., and Zaenen, I. (1978). In "Proceedings IVth International Conference Plant Pathogenic Bacteria" (Station de phytologie végétale et phytobactériologie, ed.) p. 115, Angers, I.N.R.A.

De Beuckeleer, M., Lemmers, M., De Vos, G., Willmitzer, L., Van Montagu, M., and Schell, J. (1981). Mol. Gen. Genet. 183, 283.

De Greve, H., Decraemer, H., Seurinck, J., Van Montagu, M., and Schell, J. (1981). Plasmid 6, 235.

Depicker, A., Van Montagu, M., and Schell, J. (1978). Nature 275, 150

Depicker, A., De Wilde, M., De Vos, G., De Vos, R., Van Montagu, M., and Schell, J. (1980). Plasmid 3, 193.

De Vos, G., De Beuckeleer, M., Van Montagu, M., and Schell, J. (1981). Plasmid 6, 249.

Drummond, M.H., Gordon, M.P., Nester, E.W., and Chilton, M.-D. (1977). Nature 269, 535.

Ellis, J., Kerr, A., Tempé, J., and Petit, A. (1979). Molec. Gen. Genet. 173, 263.

Engler, G., Holsters, M., Van Montagu, M., Schell, J., Hernalsteens, J.P., and Schilperoort, R.A. (1975). Molec. Gen. Genet. 138, 345.

Engler, G., Depicker, A., Maenhaut, R., Villarroel-Mandiola, R., Van Montagu, M., and Schell, J. (1981). J. Mol. Biol. 152, 183.

Firmin, J.L., and Fenwick, G.R. (1978). Nature 276, 842.

Franck, A., Guilley, H., Jonard, G., Richards, K., and Hirth, L. (1980). Cell 21, 285.

Garfinkel, D.J., and Nester, E.W. (1980) J. Bacteriol. 144, 732.

Garfinkel, D.J., Simpson, R.B., Ream, L.W., White, F.F., Gordon, M.P., and Nester, E.W. (1981). Cell 27, 143.

Gelvin, S.B., Gordon, M.P., Nester, E.W., and Aronson, A.I. (1981). Plasmid 6, 17.

Genetello, Ch., Van Larebeke, N., Holsters, M., Depicker, A., Van Montagu, M., and Schell, J. (1977). Nature 265, 561.

Goodman, R.M. (1981). J. Gen. Virol. 54, 9.

Gordon, M.P. (1979). In "Proteins and Nucleic Acids" (A. Marcus, ed.), (The Biochemistry of Plants, Vol. 6) p. 531. Academic Press, New York.

Gurley, W.B., Kemp, J.D., Albert, M.J., Sutton, D.W., and Callis, J. (1979). Proc. Natl. Acad. Sci. USA 76, 2828.

Guyon, P., Chilton, M.-D., Petit, A., and Tempé, J. (1980).
    Proc. Natl. Acad. Sci. USA 77, 2693.
Hernalsteens, J.P., De Greve, H., Van Montagu, M., and
    Schell, J. (1978). Plasmid 1, 218.
Hernalsteens, J.P., Van Vliet, F., De Beuckeleer, M.,
    Depicker, A., Engler, G., Lemmers, M., Holsters, M., Van
    Montagu, M., and Schell, J. (1980). Nature 287, 654.
Hohn, B., and Hohn, T. (1982). In "Molecular Biology of Plant
    Tumors" (G. Kahl, and J. Schell, eds.) in press. Academic
    Press, New York.
Holsters, M., Silva, B., Van Vliet, F., Genetello, C., De
    Block, M., Dhaese, P., Depicker, A., Inzé, D., Engler,
    G., Villarroel, R., Van Montagu, M., and Schell, J.
    (1980). Plasmid 3, 212.
Holsters, M., Villarroel, R., Van Montagu, M., and Schell, J.
    (1982). Mol. Gen. Genet., in press.
Howell, S.H., Walker, L.L., and Dudley, R.K. (1980) Science
    208, 1265.
Kanter, A.L., and Birshtein, B.K. (1981). Nature 293, 402.
Kerr, A., Manigault, P., and Tempé, J. (1977). Nature 265,
    560.
Klapwijk, P.M., Scheuldermon, T., and Schilperoort, R.A.
    (1978). J. Bacteriol. 136, 775.
Koekman, B.T., Ooms, G., Klapwijk, P.M., and Schilperoort,
    R.A. (1979). Plasmid 2, 347.
Leemans, J., Shaw, C., Deblaere, R., De Greve, H., Hernal-
    steens, J.P., Maes, M., Van Montagu, M., and Schell, J.
    (1981). J. Mol. Appl. Genet. 1, 149.
Leemans, J., Deblaere, R., Willmitzer, L., De Greve, H.,
    Hernalsteens, J.P., Van Montagu, M., and Schell, J.
    (1982). EMBO Journal, in press.
Lemmers, M., De Beuckeleer, M., Holsters, M., Zambryski, P.,
    Depicker, A., Hernalsteens, J.P., Van Montagu, M. and
    Schell, J. (1980). J. Mol Biol. 144, 355.
Lurquin, P. (1981). Plant Sci. Lett. 21, 31.
McPherson, J.C., Nester, E.W., and Gordon, M.P. (1980). Proc.
    Natl. Acad. Sci. USA 77, 2666.
Mendel, R.R., and Müller, A.J. (1979). Mol. Gen. Genet. 177,
    145.
Merlo, D.J., Nutter, R.C., Montoya, A.L., Garfinkel, D.J.,
    Drummond, M.H., Chilton, M.-D., Gordon, M.P., and Nester,
    E.W. (1980). Mol. Gen. Genet. 177, 637.
Ooms, G., Klapwijk, P.M., Poulis, J.A., and Schilperoort,
    R.A. (1980). J. Bacteriol. 144, 82.
Otten, L., De Greve, H., Hernalsteens, J.P., Van Montagu, M.,
    Schieder, O., Straub, J., and Schell, J. (1981). Mol.
    Gen. Genet. 183, 209.

Petit, A., Dessaux, Y., and Tempé, J. (1978a). In "Proceed-
    ings IVth International Conference Plant Pathogenic
    Bacteria" (Station de phytologie végétale et phytobacté-
    riologie, ed.) p. 143, Angers, I.N.R.A.
Petit, A., Tempé, J., Kerr, A., Holsters, M., Van Montagu, M.
    and Schell, J. (1978b). Nature 271, 570.
Schell, J. (1975). In "Genetic manipulations with plant mate-
    rials" (L. Ledoux, ed.), p. 163. Plenum Press, New York.
Schell, J., and Van Montagu, M. (1980). In "Genome Organiza-
    tion and Expression in Plants" (C.J. Leaver, ed.), p.
    453. Plenum Press, New York.
Schell, J., Van Montagu, M., De Beuckeleer, M., De Block, M.,
    Depicker, A., De Wilde, M., Engler, G., Genetello, C.,
    Hernalsteens, J.P., Holsters, M., Seurinck, J., Silva,
    B., Van Vliet, F., and Villarroel, R. (1979). Proc. R.
    Soc. Lond. B 204, 251.
Schell, J., Van Montagu, M, Holsters, M., Hernalsteens, J.P.,
    Leemans, J., De Greve, H., Willmitzer, L., Otten, L.,
    Schröder, J., and Shaw, C. (1981). In "Developmental
    Biology using purified genes" (D. Brown, and C.F. Fox,
    eds) (ICN-UCLA Symposia on Molecular and Cellular
    Biology, Vol. 23) p. 557, Academic Press, New York.
Schilperoort, R.A., Hooykaas, P.J.J., Klapwijk, P.M., Koekman,
    B.P., Nuti, M.P., Ooms, G., and Prakash, R.K. (1979). In
    "Plasmids of medical, environmental and commercial im-
    portance" (K. Timmis and A. Pühler, eds), p. 339. Else-
    vier, Amsterdam.
Schilperoort, R.A., Klapwijk, P.M., Ooms, G., and Wullems,
    G.J. (1980). In "Genetic origins of tumour cells" (F.J.
    Cleton and J.W. Simons, eds), p. 87. Martinus Nijhoff,
    The Hague.
Schröder, J., and Schröder, G. (1982). Mol. Gen. Genet., in
    press.
Schröder, J., Schröder, G., Huisman, H., Schilperoort, R.A.,
    and Schell, J. (1981a). FEBS Lett. 129, 166.
Schröder, J., Hillebrand, A., Klipp, W., Pühler, A. (1981b).
    Nucl. Acids Res. 9, 5187.
Smith, G.R., Kunes, S.M., Schultz, D.W., Taylor, A., and
    Trinan, K. (1981). Cell 24, 429.
Stahl, F.W., Stahl, M.M., Malone, R.E., and Craseman, J.M.
    (1980). Genetics 94, 235.
Tennhammer-Ekman, B., and Sköld, O. (1979). Plasmid 2, 334.
Thomashow, M.F., Nutter, R., Montoya, A.L., Gordon, M.P., and
    Nester, E.W. (1980a). Cell 19, 729.
Thomashow, M.F., Nutter, R., Postle, K., Chilton, M.-D.,
    Blattner, F.R., Powell, A., Gordon, M.P., and Nester,
    E.W. (1980b). Proc. Natl. Acad. Sci. USA 77, 6448.

Van Larebeke, N., Engler, G., Holsters, M., Van den Elsacker, S., Zaenen, I., Schilperoort, R.A., and Schell, J. (1974). Nature 252, 169.

Van Larebeke, N., Genetello, C., Schell, J., Schilperoort, R.A., Hermans, A.K., Hernalsteens, J.P., and Van Montagu, M. (1975). Nature 255, 742.

Van Montagu, M., and Schell, J. (1979). In "Plasmids of medical, environmental and commercial importance" (K. Timmis and A. Pühler, eds), p. 71. Elsevier, Amsterdam.

Van Montagu, M., Holsters, M., Zambryski, P., Hernalsteens, J.P., Depicker, A. De Beuckeleer, M., Engler, G., Lemmers, M., Willmitzer, M., and Schell, J. (1980). Proc. R. Soc. B 210, 351.

Watson, B., Currier, T.C., Gordon, M.P., Chilton, M.-D., and Nester, E.W. (1975). J. Bacteriol. 123, 255.

Willmitzer, L., Otten, L., Simons, G., Schmalenbach, W., Schröder, J., Schröder, G., Van Montagu, M., De Vos, G., and Schell, J. (1981a). Mol. Gen. Genet. 182, 255.

Willmitzer, L., Schmalenbach, W., and Schell, J. (1981b). Nucl. Acids Res. 9, 4801.

Willmitzer, L., Simons, G., and Schell, J. (1982). EMBO Journal 1, in press.

Yadav, N.S., Postle, K., Saiki, R.K., Thomashow, M.F., and Chilton, M.-D. (1980). Nature 287, 458.

Yang, F., Montoya, A.L., Merlo, D.J., Drummond, M.H., Chilton, M.-D., Nester, E.W., and Gordon, M.P. (1980). Mol. Gen. Genet. 177, 707.

Zaenen, I., Van Larebeke, N., Teuchy, H., Van Montagu, M., and Schell, J. (1974). J. Mol. Biol. 86, 109.

Zambryski, P., Holsters, M., Kruger, K., Depicker, A., Schell, J., Van Montagu, M., and Goodman, H.M. (1980). Science 209, 1385.

CHAPTER 14

RECOGNITION AND MODIFICATION OF CROP PLANT GENOTYPES USING
TECHNIQUES OF MOLECULAR BIOLOGY

Richard B. Flavell

Plant Breeding Institute
Trumpington
Cambridge, England

I. INTRODUCTION

Progress in the development of the techniques of molecular
biology over the past six years has been spectacular. These
techniques enable genes to be isolated, analysed and manipulated individually, as DNA, in a way few would have considered
possible a decade ago. Molecular biology is beginning to make
a considerable impact in plant genetics and in consequence
we are witnessing an explosion of interest in the application
of the techniques to plant breeding. In this article I want
to illustrate some of the ways molecular biology can contribute to crop improvement programs. As my title implies I will
deal with the topic in two parts. Initially I will concentrate
on how the techniques of molecular biology can assist in
genetic analysis to recognise genes and their variants and how
in favourable circumstances the techniques can be developed for
screening large populations of plants. In the second part I
will speculate on how the insertion of new genes into plants
might be useful to plant breeders.

I start with methods for assaying variation because very
often the breeder is not short of variation but finds it
difficult, time-consuming and expensive to find the particular
variant he needs. Most of the time the breeder is improving
plants by selecting complex combinations of genes of unknown
function and often individually of small effect. This complexity means that assaying the phenotype, i.e. the resultant

PLANT IMPROVEMENT
AND SOMATIC CELL GENETICS

277

interaction of all the genes, will rarely be dislodged from its
central position in plant breeding. But, occasionally, major
improvements result from the manipulation of one or a very few
genes which have a relatively large phenotypic effect. These
are the genes that the molecular biologist can help to assay.
To assay a gene the molecular biologist usually needs a
nucleic acid "probe" which is the gene, a fragment of the gene
or the mRNA gene product. He can then survey plants for the
presence of the gene or for specific variants (alleles) of
the gene. A variety of methods are available to carry out
these surveys. Some are difficult and time-consuming,
some easy and rapid. I describe below examples of these
methods which I believe will have some value, in specific
circumstances, in plant improvement programs.

II. THE MAPPING OF GENES AND DETECTION OF GROSS DIFFERENCES
     IN GENE NUMBER BY IN SITU HYBRIDISATION TO CHROMOSOMES

     Where it is desirable to know if a particular gene or
chromosome is present in a small population or to find a plant
with many more or fewer copies of a gene then the technique
of *in situ* hybridisation may be useful. In this technique
the 'probe' nucleic acid is radioactively labeled and
incubated with a preparation of cells from each plant squashed
on a glass slide, some of the cells possessing nuclei at
metaphase (Hutchinson *et al.*, 1981). During the incubation
the probe nucleic acid molecules hybridise with complementary
sequences in the chromosomes. After washing away the un-
hybridised probe molecules, the presence of the gene in the
chromosomes can be inferred by autoradiography and light
microscopy (see Hutchinson *et al.*, 1981). Its position(s)
on individual chromosomes can be mapped by studying the metaphase chromosomes.
     Hybridisation of *Secale cereale* (rye) chromosomes in a
root tip cell with radioactively labeled ribosomal RNA genes
is illustrated in Figure 1. The picture shows that the genes
are located on the short arms of one pair of chromosomes.
Furthermore, the larger number of silver grains at one of the
sites (confirmed in assays of a large number of cells)
indicates the plant is heterozygous for the number of ribosomal RNA genes clustered at the ribosomal DNA locus (Miller
et al., 1980). Thus the genes have been positioned on the
chromosomes and heterozygosity detected without making a
sexual cross and without the use of genetic markers already
located on chromosomes.

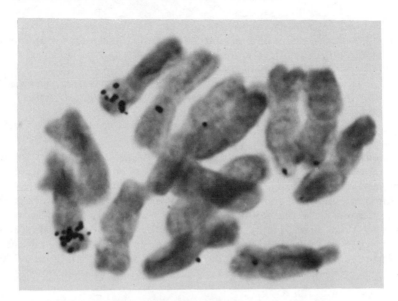

FIGURE 1. Hybridisation of ³H-labeled ribosomal RNA to chromosomes of rye, variety King II. The ³H-labeled ribosomal RNA was synthesized *in vitro* from rRNA genes purified from wheat by recombinant DNA techniques (Gerlach and Bedbrook, 1979). Further details of the experiment are given in Miller *et al*. (1980).

This example of gene localisation is easily carried out because plants contain many thousands of ribosomal RNA genes clustered at one or a few loci (Ingle *et al*., 1975). The fewer genes at a locus that are homologous to the probe nucleic acid, the more difficult it is and the longer the autoradiographic exposure necessary to detect the genes. However, genes present in only one copy have been detected in animal chromosomes (Gerhard *et al*., 1981; Harper and Saunders, 1981) and many improvements in the techniques are being developed. We can therefore look forward, with some confidence, to the time when the physical mapping of chromosomes by *in situ* hybridisation will replace conventional genetic mapping where suitable nucleic acid probes are available to the molecular biologist.

The technique of *in situ* hybridisation could be used for the routine surveying of hundreds but probably not thousands of plants. Its limitations are that useful variation rarely involves the presence/absence of a gene or gross difference in gene number. Variation in gene structure is probably the kind of variation most often exploited by plant breeders.

To assay variation in gene structure it is necessary to
purify DNA from each plant and use techniques that can reveal
changes in nucleotide sequences.

III. DETECTION OF GENETIC VARIATION USING RESTRICTION
      ENDONUCLEASES AND ELECTROPHORESIS

     Genetic variation can often be revealed by determining
the position of specific combinations of base pairs recognized
by restriction endonucleases. Several hundred such enzymes
are known (Nathans and Smith, 1975) and thus many different
combinations of base sequences can be assayed. DNA is
purified from the plant. After cutting the total plant DNA
at specific sites with one or more endonucleases the fragments
are fractionated by electrophoresis in a suitable gel.

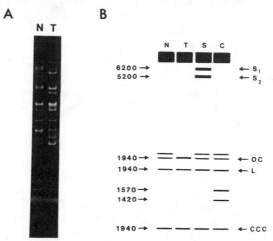

FIGURE 2.   Characterisation of maize cytoplasms using mitochondrial DNAs.   A. Maize mitochondrial DNA from plants with
N or t cytoplasms was isolated as described by Kemble et al.
(1980).   After fragmentation by the restriction endonuclease
Bam HI, the fragments were separated by electrophoresis and
detected by UV illumination after staining with ethidium
bromide.   B. Mitochondrial DNA was prepared from young
etiolated seedlings of maize plants with N, T, S or c cytoplasm.   After electrophoresis in agarose, the DNA bands were
detected by UV illumination after staining with ethidium
bromide.   OC, L and CCC refer to the open wide, linear and
covalently closed circular forms of the same plasmid. S_1 and
S_2 are linear molecules found only in S cytoplasm.   The
numbers on the left are the approximate number of base pairs
in each DNA species.   Redrawn from Kemble and Bedbrook (1980)
and Kemble et al. (1980).

With small, simple plant genomes such as those from the
mitochondria and chloroplasts, there are sufficiently few
fragments for most of them to be clearly distinguished after
staining the DNA in the gel. The separation of fragments
after digestion of mtDNA (with the enzyme Bam HI) from lines
of maize with genetically different cytoplasms is shown in
Fig. 2A. The pattern of the two lines are clearly different
and are diagnostic of the two types of cytoplasm (Levings and
Pring, 1976).

Digestion of the mtDNA with a restriction endonuclease
is not essential to distinguish these and two other (S and C)
maize cytoplasms because small pieces of DNA reside in the
mitochondria and the complement observed after electrophoresis
of the DNA is different in each cytoplasm (see Fig. 2B taken
from Kemble *et al.*, 1980). There is a linear DNA species
about 2.3 kb in mitochondria from N, S and C cytoplasms while
it is shorter in mitochondria from T cytoplasms. S mitochondria are distinguished by the presence of two linear DNA
species, S_1 and S_2, 6.2 Kbp and 5.2 Kbp long, respectively.
C mitochondria are distinguished by circular DNA molecules
containing about 1570 and 1420 bp respectively (Kemble *et al.*,
1980). These electrophoretic assays of maize cytoplasms are
rapid, taking about 2 days. This timescale contrasts with
the one or two years necessary to distinguish the cytoplasms
by crossing to tester stocks carrying the different restorer
alleles (Duvick, 1965; Beckett, 1971). Enough DNA to carry
out these assays can be isolated from a few seedlings or from
a single leaf from larger plants (Kemble, 1980).

IV. DETECTION OF GENETIC VARIATION USING RESTRICTION ENDO-
     NUCLEASES, ELECTROPHORESIS AND HYBRIDISATION TO A PROBE

In the large, complex nuclear genomes of plants, variation in individual genes cannot be detected in DNA digests
by restriction endonucleases without highlighting the genes
with a nucleic acid probe. This is achieved by transferring
the DNA fragments, after electrophoresis, from the gel to a
nitrocellulose filter, preserving the pattern of DNA fractionation achieved by the electrophoresis (Southern, 1975). Once
bound to the nitrocellulose the fragments can be incubated
with the ^{32}P-labeled nucleic acid probe and the complementary
fragments which bind to the probe are detected by autoradiography. The results of digesting DNAs from several wheat
varieties with the restriction endonucleases EcoRI and Bam HI
and incubating with a probe to highlight the ribosomal RNA
genes is shown in Fig. 3. All varieties have rDNA restriction

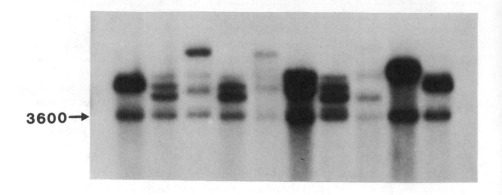

3600→

FIGURE 3.   Variation in DNA sequences around the ribosomal RNA
genes in wheat varieties revealed by restriction endonucleases.
DNAs from the different varieties were purified and restricted
with the endonucleases EcoRI and Bam HI.   After electrophoresis
(from top to bottom) in agarose the DNAs were transferred to a
sheet of nitrocellulose preserving the electrophoretic
pattern established in the gel (Southern, 1975).   The DNA was
then hybridised to ^{32}P-labelled ribosomal RNA, the nonhybrid-
ised RNA washed away and the sites of hybridisation revealed
by autoradiography on X-ray film.   The 3600 bp fragment common
to all varieties is marked.

fragments of about 3600 base pairs.   These fragments contain
sequences specifying the rRNAs.   However, there is length
variation between varieties for other fragments.   This is due
to variation in the length of the DNA which separates the
repeating units specifying the rRNAs.   Thus the use of
restriction endonucleases to fragment genes or sequences
around genes illustrates how genetic variants can be detected
without assaying the phenotype or carrying out a cross and
analysing progeny.   The drawback of the technique for large-
scale screening is that DNA needs to be highly purified from
each plant.   It is also, of course, necessary to have a probe
for the gene under study, a situation likely to be met only
rarely in the next few years.   However, if a probe is available
for a nucleotide sequence nearby to a desired allele, and there
is variation in the DNA homologous to the probe, then the
genetic linkage between the probe DNA and the allele can make
the probe useful for finding plants with the desired allele.
This approach is also useful when the desired allele at a
locus cannot be distinguished from other alleles by restriction
endonucleases.   This situation is illustrated in Fig. 4.   Here,
DNAs from two genotypes are fragmented with a restriction

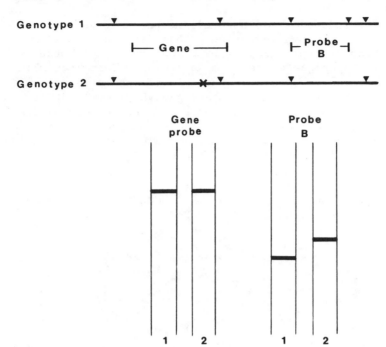

FIGURE 4.   A schematic representation of the detection of allelic variation based upon linkage of the locus to DNA sequences possessing different restriction endonuclease recognition sites.   The relevant homologous chromosomes of the 2 genotypes are illustrated above.   The restriction enzyme sites are marked by ▽. *After fractionation of the DNA, electrophoresis, transfer to a nitrocellulose sheet and probing with the gene, the two alleles (one with the mutation X) are present on fragments of identical length. However, when probed with probe B, genotype 1 can be distinguished from genotype 2 because the fragments homologous to probe B are of different lengths.*

endonuclease, fractionated by electrophoresis, transferred to nitrocellulose and probed with the gene.   The allele with the desired mutation (X) cannot be distinguished from the undesirable allele.   However, because there is a polymorphism of restriction endonuclease sites close by, the chromosome of genotype 1 can be distinguished from its homologue in genotype 2 when probe B is used.   The frequency of polymorphisms for restriction endonuclease recognition sites within a species is probably reasonably high outside the coding regions of genes, thus making allele recognition by fragment length variation after restriction endonuclease cutting an attractive possibility.

This kind of approach to allele recognition, together with the *in situ* hybridisation method described earlier, illustrates the potential value of having nucleic acid probes which map to unique sites in a karyotype. Using recombinant DNA techniques it should be possible to obtain large numbers of such probes and with painstaking work determine their linkage relationship to each other and the chromosome arms. The resulting collection of mapped probes could be of great benefit to any project where detection of genetic variation either directly or by linkage is useful.

V.   A RAPID TECHNIQUE SUITABLE FOR SCREENING POPULATIONS TO
      DETECT VARIATION IN GENE NUMBER

However advantageous are assays based on fractionating DNA (with or without restriction enzyme digestion), the need to isolate the DNA first reduces the value of the technique for screening large plant populations. Simpler ways should therefore be sought wherever possible. We (Abbot, Hutchinson and Flavell, unpublished) have recently evolved a method suitable

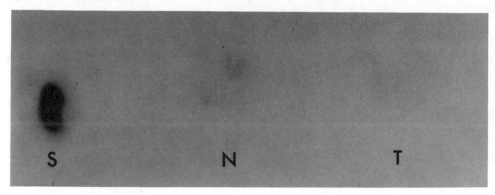

FIGURE 5.   *Recognition of S, N and T cytoplasms of maize by hybridisation of a probe to tissue segments. Segments of material from plants with S, N or T cytoplasm were squashed on to nitrocellulose, and the DNA denatured and fixed to the nitrocellulose as described in the text. After hybridisation to a ^{32}P-labelled probe of a sequence purified from S_1 by molecular cloning (Thompson et al., 1980), the extent of hybridisation was determined by autoradiography. The shapes of the N tissue segments (in pairs) but not the T tissue segments are just visible after autoradiography. The variation in the extent of the hybridisation of the probe distinguishes the mitochondrial genomes of plants with the S, N and T cytoplasms.*

for screening large plant populations of individuals which
contain different amounts of a specific nucleotide sequence.
We have applied the method to distinguish the different maize
cytoplasms. The method emerged from the characterisation of
DNA sequence differences between the mitochondria of the
different cytoplasms. Sequences in the DNA species S_1 and S_2
are not found in the mitochondrial genomes of T and C cytoplasm and are present in one copy in the N mtDNA (Thompson *et*
al., 1980; Lonsdale *et al.*, 1981). There is therefore quantitative variation in the number of copies of S_1 and S_2
sequences to distinguish S, N and T + C cytoplasms. The S_1
and S_2 sequences can therefore be used as probes to distinguish the cytoplasms. The simple assay involves squashing
the tip of a root or other small segment of tissue onto
filter paper. The cells are lysed and the DNA molecules
denatured by floating the paper on alkali. After neutralis-
ation the DNA molecules are fixed on to the paper by baking
at 80°C. The paper is then incubated in solution containing
the ^{32}P-labeled S_1 or S_2 probe and the extent of hybridisation of the probe to each of the squashed tissue segments
determined by autoradiography (see Figure 5). Hundreds of
plants can be screened in a day or two. It is therefore an
ideal kind of assay to assist in plant breeding.

The sorts of techniques described so far in this paper
present, then, new opportunities for detecting genetic variation in plants. They enable homozygous and heterozygous
lines to be recognised directly. They also enable recessive
as well as dominant genes to be distinguished once the phenotypic variation has been correlated with the nucleotide
sequence variation. Detection of single copy genes is
technically much more demanding than multicopy sequences.
When the techniques will be useful in crop improvement
programmes will depend upon two things: whether a nucleic
acid probe is available for the gene under study or a sequence
closely linked to it and whether it is expedient given the
time, difficulty, cost etc. of using these versus other
"conventional" techniques where the phenotype rather than the
genotype is assayed.

The first point, the availability of purified probes, is
dependent upon the extent to which molecular biologists, plant
breeders and geneticists get together. I cannot emphasize
this point too strongly. Breeders and geneticists must define
to molecular biologists those genes which they need to assay
but which are difficult, time-consuming or expensive to do so.
Molecular biologists must become acquainted with the needs of
breeding programmes and work with the geneticists and physiologists to uncover the means of obtaining the probes.

VI.   MODIFICATION OF PLANTS BY THE INTRODUCTION OF DNA FROM
      THE TEST TUBE

     The insertion of new chosen genes into plant chromosomes
from the test tube is not a dream for the future (Flavell,
1981; Cocking et al., 1981). It has already been achieved
by molecular biologists who have exploited the "natural"
insertion mechanism used by the bacterium Agrobacterium
tumefaciens (e.g. Hernalsteens et al., 1980). This bacterium
contains a large plasmid, a small well-defined segment of
which becomes integrated in the plant chromosome after
infection (Schell, 1979). By inserting new genes into this
DNA segment while it is still in the bacterium, and infecting
the plant cells by the manipulated bacterium, it is possible
to cotransfer the new gene into the plant chromosomes. The
plant cells which receive the bacterial DNA form a crown gall
tumor and are recognised by their ability to grow in culture
without exogenous hormones. The establishment of transformed
plants which do not form tumours (an obvious need for use in
agriculture) is now being accomplished and so molecular
biology is beginning to broaden the germ plasm of those
dicotyledonous species which can be infected by Agrobacterium.
     The current euphoria over the modification of plant
genotypes by the introduction of DNA is due also to the
impressive and rapid developments in the transformation of
yeast (Hinnen et al., 1978; Beggs, 1978; Struhl, 1979) and
animal cells (Wigler et al., 1979; Wold et al., 1979).
     The conclusions from the experiments on the transform-
ation of fungal and animal cells by incubating cells with DNA
can be summarised as follows:
1.   The frequency of cells transformed is low, even using
     single gene DNA preparations administered under the most
     favourable circumstances. Consequently, it seems
     probable that a large number of plant cells capable of
     regenerating into whole plants must be treated with DNA.
2.   For practical purposes, it is essential to be able to
     select those cells which have taken up DNA.
3.   The transformation frequency and the transforming gene
     copy number in the transformed cell depend upon whether
     the transforming DNA can replicate autonomously in the
     recipient cell or needs to be integrated into the
     chromosomes to replicate.
4.   Removal of the cell wall and incubation with DNA under
     protoplast fusion conditions is necessary for efficient
     transformation.
     Using these sorts of results for guidance many labor-
atories are now treating large populations of plant proto-

plasts with DNA vectors carrying a selectable gene which will allow the rare transformants to be selected after regeneration of cell walls, during growth of tissue culture callus and subsequent regeneration of a whole plant. Some laboratories are attempting to use dominant genes, as selectable markers, such as those conferring resistance to an antibiotic which kills plant cells, while others are attempting to convert auxotrophic plant cells into prototrophic cells by supplying the appropriate gene. It is not my intention to discuss these sorts of experiments in detail because they are in their infancy. However, they will almost certainly lead to transformed plants. Nevertheless, it is clearly important to note that this method is appropriate only for plants in which the efficiency of protoplast culture and plant regeneration is very high. This is largely limited to members of the Solanaceae.

Viruses, of course, have efficient mechanisms for infecting cells and transferring their DNA into plants. The DNA plant virus, cauliflower mosaic (double-stranded DNA) (Shepherd, 1979; Hull, 1979) and the geminii viruses (single-stranded DNA) (Haber *et al.*, 1981) are therefore being evaluated as vectors to insert additional defined genes into plants. The CaMV DNA is itself infectious and so it is not necessary to reassemble the virus *in vitro* after manipulation of its DNA. However, to date, it has not been possible to insert large pieces of new DNA into the viral genome, maintain infectivity and keep the additional DNA stably integrated in the genome. Viruses are not usually inherited through seed. Therefore the insertion of genes via viral vectors appears most suitable for the transformation of vegetatively propagated crops.

An alternative method of delivering genes into germ line cells is to inject DNA directly into pollen or egg cells. This route seems particularly attractive for those crops, e.g. cereals and legumes, which cannot be regenerated efficiently from protoplasts. Some laboratories are beginning to attempt this route. It clearly demands the ability to culture the injected cells through normal development to obtain transformed plants.

With the great interest emerging in plant transformation it is probable that several different methods of inserting DNA into plants and detecting the transformants will emerge. Different crops will demand different approaches and solutions. At this stage plant breeders should not worry about the method of getting DNA into plants. This is being achieved, or will be so, for our major crops over the next few years by many laboratories. Of more importance is what benefit this new tool can bring to plant improvement programs.

VII. USEFUL VARIATION FROM THE TEST TUBE

     To advance crop improvement by using exogenous DNA it is
essential to have in the test tube genes which have a large
beneficial phenotypic effect when transferred to the plant.
What are these genes? Single genes with large phenotypic
effects that molecular biologists are busy isolating include
those conferring herbicide resistance, disease resistance,
seed protein quality and male sterility. Other genes that it
would be useful to have in the test tube include those
altering height, vernalisation requirement, flowering time,
resistance to salt, heavy metals and drought. The list
naturally varies from crop to crop and breeding programmes to
breeding programmes, but the most impressive thing about the
lists, at the present time, is that they are relatively short!
This is partly because many characters of agronomic interest
are determined by many genes each of small effect. However,
it is also due to a serious lack of knowledge about the genetic
control of important characters. Without much more
information on the consequences of varying specific genes,
many of the potential opportunities provided by molecular
biology cannot be realised for crop improvement. The purifi-
cation of a gene by the techniques of molecular biology
demands a way of recognising the gene and a considerable
investment of skill and time. Therefore knowledge of the gene
product and the potential benefits which the gene might confer
when inserted into a crop is highly desirable before the
challenge to purify the gene is taken up. Consideration of all
these sorts of issues highlights once again the need for
close collaboration between molecular biologists, geneticists,
physiologists and plant breeders, if the tools of molecular
biology are to benefit plant improvement programmes.
     One of the very attractive features of inserting genes
from the test tube into a plant is that new genes, not found
in the species, can be used. Whether such genes will function
and their products integrate into the developmental
physiology of the new host to produce an improvement in crop
performance is, of course, a matter for experimentation.
Although the insertion of individual alien genes is a novel
opportunity, the reinsertion of alleles from the same
species is not a useless exercise because one of the great
practical benefits of being able to insert a gene from the
test tube into a plant is that *single* genes can be inserted.
No backcrossing is required to remove all the other genes
which are inserted in a sexual cross. There are no deleterious
genes tightly linked to the desired gene to make the
introduction of marginal benefit. Furthermore, by using

genes already known to function in the germ plasm of successful breeding programs we may improve greatly the chances of getting the inserted genes to be expressed correctly during plant development.

Gene expression is determined in part by the nucleotide sequences in and around the coding sequence. The sequence variation that is generated in nature may be very limited compared with what could be produced in the test tube. Therefore the insertion of a variety of different forms of a gene isolated from the same species may produce variation not presently in the hands of breeders.

Two almost inevitable consequences of introducing genes from the test tube into plants are (1) the gene will reside in different positions in different plants and (2) different numbers of copies of the gene will be stabilised in different plants, i.e. position and copy number will be very difficult to control. What novel phenotypic variation may result from altering the position of a gene or the copy number of a gene is unknown, but is is an exciting possibility that creating molecular variations of these kinds will be useful to plant breeders, even where a copy of the same gene preexisted in the plant. The inability of the experimenter to control variation of this sort emphasises that selection of the desirable phenotypic variants, after insertion of a gene, will be an important part of the use of plant transformation procedures in plant improvement programs. It also emphasises that many transformants may have to be produced and screened to ensure the production of useful variation from plant transformation programs.

VIII.   <u>CONCLUDING REMARKS</u>

The techniques for detecting and purifying individual genes are now developed, if not perfected, providing a "probe" or other means of recognising the gene is available. The ability to insert new genes efficiently into plants, including our crop species, will surely emerge over the next few years. The speed with which these techniques of molecular biology have emerged implies that they will be ready for use in breeding programs long before their potential can be realised. This is because we know very little about the genetic and physiological complexity of our crop plants and the genetic basis of yield limitations. There is therefore an urgent need to increase greatly the investment in these

areas if molecular biology is to be another tool to help
breeders maintain their momentum in providing food for the
hungry world of the next century.

REFERENCES

Beckett, J.B. (1971). *Crop Sci. 11*, 724.
Beggs, J.B. (1978). *Nature 275*, 104.
Cocking, E.C., Davey, M.R., Pental, D., and Power, J.B. (1981).
    Nature 293, 265.
Duvick, D.N. (1981). *Adv. Genet. 13*, 1.
Flavell, R.B. (1981). Proc. of Rockefeller Workshop,
    Rockefeller Foundation, New York.
Gerhard, W.L., and Bedbrook, J.R. (1979). *Nucleic Acid Res.
    7*, 1869.
Haber, S., Ikegami, M., Bajet, N.B., and Goodman, R.M. (1981).
    Nature 289, 324.
Harper, M.E., and Saunders, G.F. (1981). *Chromosoma 83*, 431.
Hernalsteens, J.P., Van Vliet, F., De Benekeleer, M.,
    Depicker, A., Engles, G., Lemouers, M., Hosters, M.,
    Ban Montagu, M., and Schell, J. (1980). *Nature 287*,
    654.
Hinnen, A., Hicks, J.B., and Fink, G.T. (1978). *Proc. Nat.
    Acad. Sci. USA 75*, 1929.
Hull, R. (1979). *In* "Nucleic Acid Research". (eds. T.C. Hall
    and J.W. Davis,) *Vol. 2*, 3. CRC Press, Boca Ratan,
    Florida.
Hutchinson, J., Flavell, R.B., and Jones, J. (1981). *In*
    "Genetic Engineering" Vol. III, (eds. J. Setlow and
    A. Hollaender) Plenum Press, New York.
Ingle, J., Timmis, J.N., and Sinclair, J. (1975). *Plant
    Physiol. 55*, 496.
Kemble, R.J. (1980). *Theoret. Appl. Genet. 57*, 97.
Kemble, R.J., and Bedbrook, J.R. (1980). *Nature 284*, 565.
Kemble, R.J., Gunn, R.E., and Flavell, R.B. (1980). *Genetics
    95*, 451.
Levings III, C.S., and Pring, D.R. (1976). *Science 193*, 158.
Lonsdale, D.M., Thompson, R.D., and Hodge, T.P. (1981).
    Nucleic Acid Res. 9, 3657.
Miller, T.E., Gerlach, W.L., and Flavell, R.B. (1980).
    Heredity 45, 377.
Nathans, D., and Smith, H.O. (1975). *Ann. Rev. Biochem. 44*,
    273.
Schell, J. (1979). *In* "Nucleic Acids in Plants". Vol. 2
    (eds. T.C. Hall, and J.W. Davies) *Vol. 2*, 195. CRC
    Press, Boca Raton, Florida.

Shepherd, R.J. (1979). *Ann. Rev. Plant. Physiol. 30,* 405.
Southern, E.M. (1975). *J. Molec. Biol. 98,* 503.
Struhl, K., Stinchcombe, D.T., Scherer, S., and Davis, R.W.
      (1979). *Proc. Nat. Acad. Sci. 76,* 1035.
Thompson, R.D., Kemble, R.J., and Flavell, R.B. (1980).
      Nucleic Acid Res. 8, 1999.
Wigler, M., Pellicer, A., Silverstein, S., Axel, R.,
      Urlaub, G., and Chasin, L. (1979). *Proc. Nat. Acad.
      Sci. USA 76,* 1373.
Wold, B., Wigler, M., Lacy, E., Maniatis, T., Silverstein, S.,
      and Axel, R. (1979). *Proc. Nat. Acad. Sci. USA 76,*
      5684.

Index

A

Acid sulfate soils
  plant growth and, 78–79
  problems of in plant varieties, 87
  reclaiming of, 79
Agrobacterium tumefaciens, Ti plasmids of,
    257–258
Albino pollen plants, 141
Alien genome combination, 99, *see also*
    Genome combination
  in alien germplasm potential combination,
    115
  fertilization in, 106
Alien genome homology, 113–114
Alien genomes, hybrid fertility and, 113
Alien germplasm
  expression of, 114–115
  utilization of, 112–115
Alkalinity tolerance, in plant families, 85–86
Amino acid analog resistance, 223–224
Amplified allozymic variability, in tomatoes,
    19–20
Androgenetic haploids, 189
Anther culture
  albino pollen plants and, 141
  cold pretreatment in, 139
  culture medium in, 137–138
  in haploid production, 131–134
  physical factors of culture in, 138–139
  pollen dimorphism in, 140
  pollen plant ploidy in, 140–141
  in various species, 135
Arthropod resistance, in tomato, 2–3, 8–9
Auxotrophic cell lines
  mutant selection and, 222–223
  in *Nicotiana plumbaginifolia,* 227–230
  success in isolation of, 234

B

Barley
  aluminum toxicity tolerance in, 87
  salinity tolerance in, 75
Boron-toxic soils, 81
Brown rust, reactions in, 59

C

Cell culture mutants, 221–234, *see also*
    Mutants
  amino acid analog resistance and,
    223–224
Cereal crop improvement, exotic germplasm
    in, 29–41
Cereals
  plant cell culture and somatic cell genetics
    of, 179–195
  somatic embryogenesis in tissue culture
    of, 183–184
Chloroplast assortment, vs. mitochondrial
    assortment, 215–217
Chloroplasts, in somatic hybrids and
    cybrids, 205–208
Chromosomal rearrangements, in
    somaclonal variation, 171
Chromosome elimination, in somatic
    hybridization, 249–250
Chromosome variation
  in F$_1$ progeny, 107–112
  karyotype, 111–112
  meiotic, 109–111
  somatic, 107–109
Clonal propagation, in plant cell culture,
    181–184
Coarse cereals, salinity tolerance in, 85
Cold pretreatment, in anther culture, 139